A
MILITARY
LIFE

An Autobiography of 24 Years of Military Service

1SGT Henry C. Wilson, Sr.

www.TrueVinePublishing.org

A Military Life
1SGT Henry C. Wilson, Sr.

Published by
True Vine Publishing Co.
810 Dominican Dr.
Nashville, TN 37228
www.TrueVinePublishing.org

Printed in the United States of America—First printing.

DEDICATION AND ACKNOWLEDGMENTS

I dedicate this book to my wife, Sarah, who stood by me during every minute of my service. I also thank our two daughters, Dimey and Dana, who made it easy for me to serve our country. I thank my son, Henry "Clarence" Wilson Jr., for suggesting that I take time to record these events that I have spoken about constantly over the years. If it had not been for him, I would not have recorded these memories of my military experiences and my life experiences before and after serving our country.

I dedicate this book in memory of my dear friend Gary Dellinger, with whom I served in the United States Army Recruiting Command for many years. After I decided to write this book, I received word from Gary's wife, Renee, that he had passed away. God bless Gary's family, and I pray that my memories of him remain fresh in my heart.

I also dedicate this book to Leotis Dixon, who passed away in January 2025. He was a veteran who served in the United States Marine Corps. Our families met in Amarillo, TX, and Leotis and I became best friends very quickly. We never shared a dull moment. I pray that God will be with his family forever.

Writing this book was extremely therapeutic for me, though at times upsetting. Sometimes I laughed at the funny events. At other times, I teared up, and at times I became very angry as I recalled difficult experiences while writing. Thank you, son. Thank you, Family. I love each of you and your immediate family members dearly.

TABLE OF CONTENTS

DEDICATION AND ACKNOWLEDGMENTS 3

CHAPTER 1 9
It All Began with A U.S. Army Draft Notice

CHAPTER 2 19
Basic Training (Discrimination During Basic)

CHAPTER 3 25
My Advanced Individual Training (AIT)

CHAPTER 4 28
My Vietnam Tour

CHAPTER 5 31
Our Arrival in Vietnam

CHAPTER 6 34
My First Combat Experience

CHAPTER 7 39
My New Assignment to the 539th Transportation Bn

CHAPTER 8 43
My Mess Hall Leadership Experience

CHAPTER 9 46
We Are Really All the Same

CHAPTER 10 50

My Vacation Home / Dr. Martin Luther King Jr. Assassination

CHAPTER 11 52

My First Stateside Assignment

CHAPTER 12 65

The New Mess Sergeant

CHAPTER 13 71

My Assignment to The General's Staff

CHAPTER 14 86

My Departure from the General's Staff

CHAPTER 15 88

Drill Sergeant Training

CHAPTER 16 92

My First Drill Sergeant Assignment

CHAPTER 17 101

General Coleman's Departure from Fort Jackson

CHAPTER 18 105

I Got a Break

CHAPTER 19 112

My First Correctional Specialist Assignment

CHAPTER 20 131

My U.S. Army Recruiting Experience

CHAPTER 21 .... **159**

Community Service

CHAPTER 22 .... **162**

Recruiting Awards and Progress

CHAPTER 23 ....**172**

Career Counseling Course Preparation

CHAPTER 24 ....**177**

Counseling Duties at the Military Entrance and Processing Station (MEPS)

CHAPTER 25 ....**181**

My Promotion to Master Sergeant

CHAPTER 26 .... **184**

Senior Career Counselor, Amarillo, Texas

CHAPTER 27 .... **195**

My Assignment as First Sergeant of the Amarillo Recruiting Company

CHAPTER 28 ....**206**

The Company Commander Arrives

CHAPTER 29 ....**211**

One Year as First Sergeant

CHAPTER 30 .... **215**

Amarillo, the Company of the Year

CHAPTER 31 .... **218**

Sarah's Selection as Tax Auditor Manager

CHAPTER 32 221

Changes within the Battalion Leadership

CHAPTER 33 225

My Departure from my Amarillo Assignment

CHAPTER 34 238

Preparation for the United States Army Sergeants Major
Academy (USASMA)

CHAPTER 35 240

The United States Army Sergeants Major Academy
(USASMA)

CHAPTER 36 245

Class Starts

CHAPTER 37 252

The Military Science Phase

CHAPTER 38 256

Graduation and Preparation for My Next Assignment

CHAPTER 39 258

My Assignment at Fort Riley, Kansas

CHAPTER 40 280

The Unit Provisional Equipment Custodial Organization

CHAPTER 41 283

My Retirement

SUMMARY 287

CHAPTER 1

It All Began with A U.S. Army Draft Notice

1963 through 1965 were transformative years of my life. During this time, a significant number of guys that I hung out with enlisted in the U.S. Army. To no avail, a few of them tried to convince me to join the Army prior to my graduating from high school. I had absolutely no intention of dropping out of high school. I didn't give them a second thought. After graduating from high school, I went on to enroll in Tennessee State University, where I attended briefly. Then it happened.

One evening, after just being hired by Sears, Roebuck and Co., I returned home to find my mother and several siblings sitting around with long and, what appeared to be, worried faces. My mom handed me a white envelope with a letter that read, "Greetings," and other words that I really don't remember. Factually, I was ordered to report to the Memphis, Tennessee, induction station on September 15, 1966.

Since my family knew what the letter was for, my response simply confused them. I imagined they thought I had lost my mind. It was simply joy and excitement that caused my surprising response. I guess I was just happy to get away, which ultimately turned out to be a great thing for me.

I was sworn into the United States Army on September 16, 1966. Why not the 15th, you may ask? Well, this was one of those times when I believe God guided my life. I believe you will understand my meaning as you continue reading my story.

Military Entrance and Processing Station

On September 15th, hundreds of qualified men were processed and sworn into the U.S. Army, except me. Why not me? Well, all the inductees were instructed by the sergeant who conducted the briefing: "Don't do anything or go anywhere until you are told." So, I just sat there alone waiting for my name to be called. After a while, I saw a soldier coming toward me with a strange look on his face.

"Why are you still here?" he asked.

"No one called my name," I answered.

As it turned out, the explanation was that the doctor who conducted the physicals, for some reason, placed my medical folder in his desk drawer and left the induction station. Therefore, I was held over until the next day, September 16, 1966.

I was given a room for the night in the hotel in downtown Memphis that housed those who were being processed for the military. This was no problem for me. In fact, the oversight gave me a chance to visit Sarah Harbin, the young lady I was dating and who is now my wife of 57 years, one last time before departing for basic training. Needless to say, I was thrilled by the oversight.

On September 16, 1966, I was sworn into the U.S. Army and immediately departed Memphis for Fort Campbell, KY, where I attended basic training. During in-processing, I received three full sets of U.S. Army uniforms, which included work uniforms, summer and winter dress uniforms with shirts and ties, and more underwear than I had seen in my entire life. I was extremely excited to get all those clothes.

You may ask yourself, "Why was he so excited?" You must understand, when you grow up "PO" (which means real poor in my dialect) like I did, you don't get a full set of anything, which includes a full meal. A lot of my clothes were "hand-me-downs or borrowed." So now you're wondering, if I was so poor, how did I afford to attend college? To be honest, I have no idea.

All right, let me explain, and maybe it will help you understand my college enrollment a little better. All over this country, it is commonplace to hear about teachers molesting students, and to be real with you, some teachers have been molesting students for years. I know, because I was molested during my senior year of high school by two of my teachers, and had sexual conversations with another one by telephone.

One of the two teachers forcibly had sex with me once and would have done it again if I hadn't been able to fight the teacher off the second time. The other teacher lured me to her apartment with the pretense of helping me study for my college entrance exam.

Initially, her plan didn't work because I was very naïve. You see, I told two of my friends about this teacher who I thought was helping me get into college. Both classmates thought it would be a good thing for them, also, because they were planning to attend college as well. The three of us went to the teacher's apartment, where we just sat and talked. After our first visit, the teacher approached me in the school hallway the following week and whispered, "Are you coming down tonight?"

I responded, "Yes, ma'am, if you want me to."

She moved closer to me and whispered, "Yes, I do, but come alone this time. We can't get anything done with too many people there."

At this point, I didn't know her true intention was clandestine. As instructed, I met the teacher at her apartment. As I stepped into the apartment, she handed me a glass of Teacher's Scotch mixed drink. Although I was completely surprised, I still wasn't suspicious of anything. As I entered the living room area, I noticed that the lights were turned down very low.

She had also placed a portable record player on the floor with several records placed neatly around it. Although telltale signs were all around, I didn't suspect anything. Poor me. (Smile.) As I mentioned earlier, I was naïve and kind of ignorant about life back then, and, based on our news media, some of the same things are happening today in our schools because of a few sick and dishonest educators.

After a few moments, she put on a slow record and asked me to dance. At that point, I could feel the buzz from the liquor I had consumed. As we danced slowly, my knees began to weaken, and my body shook with nervousness as I thought,

What's going on here? Here I am with one of my teachers, almost three times my age, holding on to me like a girl my age. Hell, I used to think teachers didn't even shit. Yes, that's right. They were like very perfect people, like gods, to me.

Then it happened. As we slowly danced, she looked up at me and said, "Why do you think I asked you to come to see me, child?" That's what she called me—child.

I nervously responded, "So you could help me study for the exam to get into college?"

"No, child," she said, "I'm in love with you."

I thought I was going to faint or at least shit my pants. All I could remember at that point was that I left the apartment like a bat out of hell. I remember driving north on Highway 51 like I was on a racetrack. Although I took control of my senses and slowed down, my heart rate continued to race out of control all the way home.

During the next week, I tried to avoid that teacher. Even during her class, I couldn't look at her. It was class as usual until the following Friday. Then, shortly after class, *Oh shit, here she comes!* I thought to myself. The teacher cornered me as I stood alone. It really seemed like she was waiting for this opportunity. She leaned close to me and whispered, "Are you coming to see me tonight?"

I was so shocked that she would approach me after my response to her advances the last time. I hesitated for a moment, which prompted her to insist that I come see her. "It will be ok, child," she said.

I visited this teacher several times during the following months; however, I was never comfortable. She was very intimidating.

On one occasion, as I approached her apartment, there was a female student I recognized entering the apartment of a teacher who lived in the same apartment complex. Of course, I mentioned this to the teacher I was visiting. "Oh, child, no problem, that girl is her lover."

Two women, I thought in shock.

"Let me tell you something, child," she responded. "There's a lot of that happening at school. There are several teachers who want you. That's all I hear in the teacher's lounge: Henry Wilson this, Henry Wilson that," she sarcastically shouted. She looked me in my eyes with a devilish grin and stated, "I got you now, so they can just back off, right?"

Nervously, I whimpered, "Yes, ma'am." She was very aggressive and demanded that I be her man.

During my high school commencement program, as our class prepared for graduation, she came up to me as I stood at the back of the line outside and whispered, "Just think, in a few minutes we will be able to tell the whole world all about our relationship."

That made me so nervous that I could have messed up my pants. I really don't remember anything about the commencement after that, except that I could feel my insides acting up during the entire program.

Anyway, that teacher is how I got into college. I have no idea what she did to get me enrolled, but it happened. All I know is she told me, "You are going to college at Tennessee State University (TSU)."

Hell, the only reason I knew anything about TSU was because she had me take her to Nashville when our basketball team made it to the state basketball tournament. I never took a test or paid fees or anything else. All I knew was when and where to go to register for my classes.

That same teacher also gave me a fully furnished house in Nashville. One day, she gave me a tour of the house and told me, "Child, this is yours. You don't have to do anything but be my man."

Naturally, she paid the bills associated with the house. She would also travel from Memphis to Nashville every two weeks for sex and to replenish my groceries. She seemed to get pleasure from taking care of me, which included taking me to a corner eatery at Edgehill & 12th Ave. just to show me off. Embarrassingly, she would go out of her way to introduce me as her man. We would sit, drink beer, and eat.

Women who seemed to be her age (about the same age as my mom at that time) would stop by our table, mostly at the urging of my teacher, smiling and talking about stuff I couldn't care less about. The fact was, I didn't want to be there.

After about five months of school, I just couldn't take it anymore. College wasn't for me. I went to a few classes for a few weeks. The rest of my time was spent in the Student Union Building (SUB) playing Bid Whist, a common card game at that time. That year, I can say I graduated Magna Cum Laude in Bid Whist, which unfortunately doesn't even count for kindergarten graduation. Get my drift? (Smile.)

On top of not wanting to be in school, I still had to deal with the teacher. Whenever she visited Nashville, she would first take me into a small bedroom where we would have sex. After that, she would be direct and allude to what she wanted me to do for her in the courtship. This went on until January of 1966, when she came to visit and jumped all in my ass about something she overheard two of my classmates saying about me. I mean, she was like a rabid dog. She exploded on my ass. She was pissed, and it reminded me of the first time I saw her that way.

It was during the Christmas of 1965 when she made me go to a Catholic Church service with her. At the end of the program, she asked me how I liked the service. To this day, my motto is, "If you don't want to know the truth, don't ask me." I told her that I didn't like the service because I didn't understand what was going on.

That was the first time I saw her pissed. Now here she was again, pissed at me, digging in my ass about something she heard two female classmates talking about, which was true. She asked me about the Christmas present I gave Sarah, the lady I was dating, and I didn't lie. "Yes, ma'am, I gave Sarah a Christmas gift." I cannot describe the expression on her face.

Later that night, I could hear her and a man in the other room talking, which woke me up from a light sleep. Now I'm getting really scared because I have never been in a predicament like this. They were apparently drinking, laughing, and joking loudly. It was as if they deliberately wanted me to hear them, which was, I'm sure, all a plan by the teacher to scare me.

It worked. After a while, I heard the bedroom door open. In a demanding and boastful voice, she shouted, "Wake up and let me introduce you to a real man."

All I know is I spoke to the guy. I don't know what happened the rest of the night. I guess I blocked everything out. After that encounter, I grew increasingly afraid of this teacher. On top of that, I was really concerned because I was deceiving the woman I really wanted to be with.

The following morning, the teacher came into my room with a pistol in her hand. "You're scared, aren't you?"

"Yes, ma'am," I said as I felt my insides acting up.

"Hell, after the way you have treated me, you ought to be scared."

What the fuck are you talking about? I'm the victim here. Of course, I thought that to myself. I just sat there thinking, I can't do this anymore. So, I refused to yield to her demands anymore. After I refused to continue to live under her lifestyle, you know, being her man, she stated, "You are on your own then."

A relief came over me that I had never felt before. I wanted to shout, but I didn't. She looked really sad, but I didn't care, because she had brought it all upon herself. After that, all the support, funding, room, and board she was providing were suddenly cut off.

Please understand and believe me when I tell you that we had some caring and wonderful teachers at Woodstock High School. I had more than one teacher talk with me about life and its expectations when they didn't have to. Yes, I also received counseling via the paddle called "The Board of Education," which helped me think more clearly. (Smile.)

Please do not judge the entire teaching staff by a few unethical teachers. The fact is that a great number of our teachers were responsible for educating students from within the Memphis Board of Education and the Shelby County Board of Education Systems. During recent years, I learned that several of my classmates became teachers, school principals, and other education administrators within Memphis and Shelby County, Tennessee.

Back to my college experience.

I really hated to leave Sarah, but with no place to live, I was forced to return home to Millington, TN. I visited when I could during the rest of the school year, and when she came home for the summer of 1966, we spent lots of time together. Then it happened. Around midsummer, I received my draft notice.

CHAPTER 2

Basic Training
(Discrimination During Basic)

Boy, did I look good in my U.S. Army uniform. It reminded me of my Boy Scout uniform, which was the first uniform I ever wore. I was very proud to wear that scouting uniform. I was told that I looked very handsome when wearing it.

Actually, I worked hard to earn the money just to purchase my uniform. Unfortunately, I made the mistake of telling my mom what I wanted to do with it. She took the money, telling me that she needed it to buy food for the family. At 14 years old, buying food was not on my radar.

I cried so hard and so long that my eyes were crossing, and tears and snot were meeting under my chin. I guess Mom thought I was losing my mind and ended up giving me my money back. I don't know how I got there, but I didn't hesitate to go straight to the store to buy that Boy Scout uniform. Thinking back now, it's kind of funny, but at that moment, nothing about it was funny.

Ok, back to my basic training.

During basic training, I was a member of a platoon that had three drill sergeants. The senior drill sergeant was white, with two assistant drill sergeants: one white and one colored (as described then). The reason I point out the race of each drill sergeant is because it is very important to my story. You see, although I didn't realize it until many years later, this is where I experienced racial discrimination in the Army for the first time.

My senior drill sergeant introduced himself and his two assistants to us and proceeded to ask if any of us had Reserve Officer Training Corps (ROTC) training in high school or college. Because I served a brief stint in the Air Force ROTC at TSU, I raised my hand. He proceeded to assign me as the squad leader of the second squad.

Each platoon had five trainee leaders: four squad leaders and one platoon leader. I didn't know anything about what the other squad leaders were doing or how well they were doing. I just had my soldiers complete the tasks our drill sergeants assigned us to do each day.

How I managed my men must have worked, because after about four weeks of training, the senior drill sergeant came into our barracks (our living quarters), raising all kinds of hell. He proceeded to relieve all the trainee leaders from their positions, except me. Honestly speaking, I was confused about what was happening until the senior drill sergeant called me into his office and told me that he was very pleased with how I managed my men and that I was doing a great job. He went on to tell me to select someone from my squad to be the platoon leader and to make sure I taught the selectee how to do the job. "Yes, Drill Sergeant," I stated.

Right away, I selected a very well-mannered trainee to serve as platoon leader. It was years later, when I was telling this story to a friend that I grew up with in Millington, TN, that I realized what happened. I thought to myself, Oh shit, wait a minute, I got screwed!

Ok, let me explain. There were four squads within each platoon. With this change, I became subordinate to the new platoon leader whom I was directed to select and train. You see, all members of my squad were white. Get it? That meant that anyone I chose would create a white platoon leader.

I proceeded to do what I was told with pleasure, not realizing that the senior drill sergeant's direction was discriminatory. Since I became the senior trainee leader and the best qualified, I should have been placed in the platoon leader position. There's more, but I'll explain later. Keep reading.

I did extremely well during basic training, including training the new trainee platoon leader (aka the white guy) to do the job that I realized years later should have been mine. I did exceptionally well on the final physical training test of basic training. I mean, I smoked that sucker! There were five events: the inverted crawl, horizontal ladder, grenade throw, run-dodge-and-jump, and the last being the one-mile run. You were scored in each event based on the total points you received during the time allotted for that event. The faster you completed the event, the higher your score.

The grenade throw was scored based on the number of the circle that your grenade landed in. The target had several circles, with the smallest circle being the bullseye (in the center), which held the highest score. To put the size of the circle into perspective, the largest circle, and I am guessing, was about eight feet, measuring from bottom to top and the same from left to right. All I remember is that from approximately 20 meters, I consistently threw the grenade into or close to that little circle.

After the completion of the fourth event, the colored (black) drill sergeant whispered, in confidence, that I had one trainee to beat in the one-mile run to become the trainee of the cycle in physical training. I appreciated him telling me that, but I had no idea what that meant, and it really didn't matter because I always did my best in whatever I was involved in. In fact, let me tell you about my first track competition in high school that may help you understand what I mean.

In high school, each year we had an activity day called May Day. There were many activities, but the ones I specifically remember included wrapping the Maypole and track events. Anyway, I participated in wrapping the Maypole, and I was chosen to compete in the 7th, 8th, and 9th-grade 220-yard outdoor track races. I knew nothing about the layout of the track back then except where I was supposed to start and finish the race.

I remember the most exciting part of the event was the trash-talking by the upperclassmen who were my competition. They were telling me how badly they were going to beat me in this race. I remember just smiling at what they were saying, not knowing what I was going to do or how I was going to do it. Then the Physical Education (P.E.) Teacher put each participant in place.

When he placed me behind everybody else, in the inside lane, I kind of got upset because I didn't understand. But the race was about to start, so I said, forget it. Here I was behind everybody from the start, and I was thinking, *There is no way I can win this race, but here I am.* I knew to run as fast as I could when the teacher fired the shot to start the race.

Remember, I'm thinking there is no way I can win this race from where I started. I told you I was ignorant of a lot of things back then, and this was another one of them. I didn't know why they put me way back there in the first place. It seemed so unfair, but I know now. Just keep on reading.

When the race started, these guys seemed to be forever in front of me. Then I noticed that as we approached the turn going into the top of the track, I had caught up. As I entered the top of the turn, I began to pass the other runners. At that point, I clearly noticed that I was much faster than they were.

When the other runners topped the turn, I was headed to the finish line. The P.E. teacher, who was timing the race, was at the finish line waving at me to slow down because I already had the other runners beat. It was as if I had wings in my feet and a jet in my ass. I was flying!

I still didn't realize how fast I was until the high school coach approached me to start practicing for the school track team. I believe I practiced with one of the senior one-mile runners. We practiced running and conditioning after school together two or three times. That's when I realized that track wasn't for me.

Ok, back to my basic training test.

When I finished the one-mile run, the white trainee was just rounding the last turn behind me on the quarter-mile track. Even though I finished first, the white trainee was appointed the Trainee of the Cycle and got the trophy.

Out of all the 25 trainee leaders, a prior service Marine and I were the only trainee leaders in the entire company, so I was told, who were not removed for not doing their job effectively. That says a lot for me, because that Marine had been through the training before. Each of the white trainee leaders, including the one I trained as the trainee platoon leader, was automatically promoted to the grade of Private/E2, just like the regulations dictated. Did I get promoted? No!

If you are keeping count, that makes three significant experiences of racial discrimination that I encountered in a very short time while in basic training with the U.S. Army. No problem. As it turned out, I believe my promotions came faster than most soldiers in the same career field. Don't worry, I'll tell you all about my grade advances. Keep reading.

CHAPTER 3

My Advanced Individual Training (AIT)

After basic training, three other trainees and I were the only graduates who did not get assigned to a Combat Arms Career Field. In other words, out of 1,100 trainees, only four of us were assigned to support jobs. Two of us were assigned as Food Service Specialists, and the other two were assigned as Medics. If you remember, I was drafted into the Army, which is a two-year commitment.

However, during my in-processing at the induction station, I volunteered for the U.S. Army's food service program. As collateral for the training, I had to extend my term from two to three years, which changed my status from the United States (US) Army to the Regular Army (RA). Why is that terminology important? Well, I didn't know then, and the fact is, I really didn't care. However, ultimately, I learned that my decision to volunteer to work in food service was an advantage for me.

I later learned that the US status is assigned to all soldiers who are drafted into the U.S. Army, and the RA is assigned to soldiers who volunteered to serve. When I volunteered to accept the food service training, my status changed to RA.

My Advanced Individual Training was uneventful. All I can say is that I was a trainee leader during this phase of training (Cook's and Baker's School) as well. I proved to be a great cook. At least that's what people would tell me throughout my food service assignments. Keep reading, and you will understand what I mean.

After graduation, two other graduates, who were white, and I were sent by train to Fort Benning, Georgia, for our first permanent assignment. Upon arrival at personnel, the in-processing clerk, who was white, looked at us and proceeded to assign us to our first permanent assignment. The two white soldiers were assigned to work at Fort Benning in a local unit. The clerk looked up at me, shook his head, and said, "You are assigned to the 520th Transportation Battalion Headquarters Company. They are deploying to Vietnam as we speak. You have three weeks."

This was strictly perception on my part, but this sounded like a race thing to me. Remember, I was young and naïve; therefore, this was no big deal to me back then. The fact was, it didn't matter anyway. Part of the unit had already departed for Vietnam. I was assigned to the unit's Headquarters Company Mess Hall Staff. We, along with other company support staff, were scheduled to depart for Vietnam in three weeks.

Once I arrived at my work site, I was introduced to the staff that I would be working with. Within two days, I learned that the two youngest cooks were very competitive. We were responsible for feeding the headquarters company soldiers. I didn't do much cooking before I left on a two-week leave prior to leaving for Vietnam. However, there was an initiative that they saw a need to impose upon me. Let me explain.

The shift leader sent me to about five or six mess halls in the battalion to borrow a "liver skinner." Now, if I asked my grandson, who is seven years old, what a liver skinner is, he would probably know that there is no such thing. Not only did they trick me that time, but they got me twice. The second time, the shift leader sent me to borrow a meat stretcher. However, the joke was on them because I had no idea that there was no such thing as a "meat stretcher" or "liver skinner."

Years later, after I realized this, I just laughed and laughed. Actually, I thought what they did was funny, and they must have thought, *Man, that dude is really dumb. He is not going to make it in this man's Army.* I get tickled every time I think about it.

CHAPTER 4

My Vietnam Tour

While on my pre-Vietnam leave, I spent lots of time with Sarah, the only young lady that I've ever loved. We spent the time talking about marriage, so we got engaged and set our wedding date for June 1968.

In March of 1967, I returned to my unit at Fort Benning, Georgia, where we boarded an airplane to Oakland, California. Then we boarded a troop transport ship called the USS *Wiggle* for our 21-day journey to Vietnam. We set sail immediately after everyone was on board.

My experience aboard the ship was ok. I learned quickly that the best thing for me was to go to the deck and get lots of fresh air first thing in the morning and several times during each day. This, I believe, is what kept me from getting seasick, because there were awful storms during the first five days of the trip. The waves were so big that it seemed as if the ship was going to tip over at any time.

Guys were lying in their own vomit all over the deck. Unfortunately, there was nothing anyone could do for them.

The men just stayed where they were until their seasickness wore off. I, of course, did great. No problems at all. Quick story.

On a recent visit to the VA Hospital's Dental Clinic in Nashville, TN, a fellow veteran saw my cap with the Vietnam Veteran logo and, of course, he spoke, saying, "Welcome home, brother."

Immediately, we began to talk about our military experience. During our conversation, we learned that we traveled to Vietnam at the same time on the USS *Wiggle*. We reminisced about some of our same experiences. I also experienced another pleasurable moment where a white Vietnam Veteran greeted me as "Brother." The visit was very enjoyable and interesting.

Due to the number of cooks on the ship, I only worked about four hours during the entire trip. I was assigned as a fry cook. I cooked what seemed like a ton of shrimp. I also did a lot of thinking about what to expect in Vietnam. Of course, I missed my mom and siblings, who I lived with back home, and my fiancée, who stayed on my mind a lot.

The most exciting thing that happened during the trip was watching the dolphins and whales swimming alongside the ship. They were so beautiful, alternating back and forth for miles. I often thought about how God's creatures seem to do things so naturally that we don't expect of them.

During our trip, we had a two-day layover in the Philippines, where another exciting thing happened. The soldiers in pay grade E5 (sergeants and above) were the only soldiers allowed to leave the ship to go into the city. All the rest of us (E4s and below) had to stay on the ship and do whatever came naturally: play cards, write letters, etc. After about 12 hours, the soldiers started returning to the ship. That's when the fun began. Ha, ha, ha.

These guys started telling us about their experiences, including sex or lack thereof. Some said they paid for a woman to have sex with them, but got a guy instead, and didn't know it until the so-called fun was over. There were some who got all their money taken and had no fun at all. We, the lower-ranking soldiers, poked fun at them for the next 24 hours. Personally, I kind of felt sorry for them. You see, the leadership aboard the ship decided that the E4s and below soldiers were not mature enough to leave the ship. Yeah, right. (Smile.) Anyway, the next morning when I woke up, we had already departed the Philippines on our last leg to Vietnam.

CHAPTER 5

Our Arrival in Vietnam

Everything got serious once we docked in Cam Ranh Bay, Vietnam, where we were met at the ship by a transportation unit driving water-to-land vehicles. The soldiers assigned to the headquarters company, including the Company Commander, Administrative Staff, etc., were in the first group to offload. The most disturbing part of our exodus from the ship was when the noncommissioned officer in charge of troop movement made this statement to our senior officers: "Welcome to Vietnam, sir. Glad you all made it safely. The sad part is that half of these soldiers will not make it back home."

That was the most unprofessional and disturbing thing a senior noncommissioned officer should say in a situation such as the one we were in. I heard the statement, and I'm sure most of the other soldiers heard him as well, especially those in the front of the boat. Fortunately, I didn't let it bother me. To me, the flight from the landing site to our duty station was the worst thing during the trip. I was afraid something awful would happen, you know, like the airplane going down.

It was only my third flight ever, but that C-130 airplane didn't compare to the jet that flew me from Nashville to Richmond, VA, after basic training. This flight took only about 30 minutes, but that C-130 shook and rattled the entire trip. The expressions on most, if not all, of those senior soldiers' faces added to my fears. I thought, *These guys have been in service for years, and they look more afraid of flying than I am.*

Once we arrived at the company area, I was surprised but pleased to see the area I would be living in for the next year. The engineers had built a great company area consisting of motel-style living quarters and our work areas. The company area also consisted of the battalion command bunker, troop bunkers, and other necessary buildings. The mess hall, which was my work area, was well built with a fully equipped kitchen and a beautiful dining room. The Battalion Commander had a special dining area equipped with a call button to summon the Mess Sergeant when needed.

Quick story: The Meatballs

I remember once, after the Battalion Commander (everybody called him CJ) had been served his lunch, we heard the buzzer in the kitchen go off. All of us went, "Oh shit," at the same time because we didn't know what to expect. The Mess Sergeant went running to the Commander's table to see what he wanted. When he returned to the kitchen, the three of us just stood in place in wonderment, watching as the Mess Sergeant went directly to the pan of Swedish Meatballs, took out two meatballs, put them in a bowl, covered them, and put the bowl on top of the stove. He turned to us and said, "Don't mess with these meatballs. The Battalion Commander wants the same thing for supper. He said, "These are delicious."

My two counterparts, the very competitive cooks, looked at me with complete envy. You see, I made those meatballs. I just smiled and went about my work. I was the least experienced cook in the mess hall, so this was a big thing for me. My cooking abilities, attention to detail, and willingness to do whatever it took, honorably, to get the job done helped me receive promotions much faster than soldiers in the same career field. Keep reading. I'll explain later.

I never told anybody that I changed the recipe just a little to get that delicious taste. Obviously, the Colonel's palate received it well (Smile.) I'll share more of my on and off-duty cooking experiences with you soon. Keep reading.

CHAPTER 6

My First Combat Experience

My first combat experience was an enemy rocket and mortar attack. During a battalion muster earlier in the day before the attack, the Battalion Commander informed us that he had received information from the Intelligence Unit that enemy activity had been spotted in our area. He also informed us that he was expecting an enemy attack during the night. Each of us was given our assignment if the attack occurred. I was assigned to ammunition distribution.

I was not a combat soldier, but I knew that was not the safest job to have. Then again, no job is safe when you are in a combat zone. The best part was that I was assigned to the command bunker, and it was well fortified.

The soldiers were instructed to stack our uniforms and weapons in a pile next to our beds so that we would be ready for action at a moment's notice. Our intelligence was spot on. The Viet Cong attacked us with rockets and mortars (bombs) around midnight. I was awakened by the sound of sirens and bombs going off all around us. I could hear small weapons fire from the perimeter guards as well. It sounded like they were all around us. Boy, was I afraid. I just grabbed all my clothes and equipment, or so I thought.

When I arrived in the command bunker and started to get combat-ready, I realized that I had left one of my boots back in my room. Immediately after the bombing stopped, I rushed back and got my boot. Just as I returned to the bunker, our soldiers were ordered to take our assigned combat positions in case of an enemy ground attack.

I was positioned inside our company area about one hundred meters from the base perimeter line. I was in full combat gear, including my M16 rifle with a fixed bayonet, combat flak vest, grenades, and a full load of ammunition. As everyone proceeded to our assigned position, I remember hoping the enemy didn't execute a ground attack. Thank God I didn't have to haul that fricking ammo.

The only other close call I witnessed was when my friend and I were relaxing one Sunday morning on the outer perimeter of the base in our two-and-one-half-ton truck. We were just chilling, watching the Military Police (MPs) as they searched a group of Vietnamese who were trying to enter the base.

There were two things my friend, who was from Houston, TX, really loved to do: reading and smoking pot. Seriously. (Smile.) He took the Sunday paper, and I put some drinks on ice in a Mermite container (military term for cooler). We took the food truck and drove around the base perimeter road until we saw the MP checkpoint. We decided to park on the side of the road about 250 meters from the checkpoint, facing the MPs. We watched the action and talked about why we were in Vietnam and our hopes of going back home to the World, as we called America.

My friend lit a marijuana cigarette and started to read the paper. I opened a cold soda and started to read the comics. About 30 minutes after parking, we heard shots coming from the field covered with tall vegetation to our left. Just as we heard the gunfire, I noticed the MP tactical vehicles turn and face their weapons in the direction of the sound of the gunfire.

We could hear rounds of bullets going past us, so we took cover on the passenger side of the truck. Being the driver, and with everything in the cab, I had to get out on the driver's side of the truck and go around to the passenger side to take cover behind the large truck wheels. You should have seen me trying to cover my ass and my head at the same time.

It was funny after everything was all over, except when I found out that a soldier from a company motor pool (a mechanical maintenance company) was killed during that attack. He was farther inside the compound from where we were and did not seem to be in the line of fire—well, sort of not in the line of fire.

But it was kind of freaky that the bullets passed us and hit that soldier. Who am I to say why things happen the way they do? I felt sorry for the other soldier. I thank God that my friend and I didn't get hit during the enemy attack. Ok, back to the Kitchen.

During the next few months, I was very easily the best of the three cooks. After about six months in Vietnam, the Battalion Commander was ordered to transfer several of his troops to another company within the Battalion so that all troops of both companies would not rotate back to the World (United States) at the same time.

I was one of those transferred. When I was informed that I would be reassigned to another company, I became concerned that it would take me longer to advance to the next grade. Although doing the best job possible as a food service specialist, I often wondered what it would be like to be a sergeant (SGT). That motivation, together with the fact that I was constantly told that I was a great cook, prompted me to take advantage of a statement that the Company Commander made when he told me that he hated to lose me.

"Thank you, sir," I responded. "I have really enjoyed working here. I am sure I will enjoy working in the 539th Company Mess Hall. I am just concerned that it will take me longer to get the next grade. Do you think there is any way I could be promoted before I am transferred?"

"You have done a great job, Henry," the Company Commander stated. "I have heard nothing but great things about you. Let me check the books. If we got the cooks' allocations for an E4, you got it. I'll let you know."

I thanked the commander and returned to the kitchen. Shortly after our conversation, the Battalion Clerk, who processes all allocations for promotion, came into the Mess Hall. The sergeant and I talked all the time, so I quickly took the opportunity to share with him my conversation with the Company Commander and ask about the allocation. He told me, "I will check and get right back to you."

That he did, with great news. The allocation was there, and I was advanced to Specialist 4th Class (SP4) within days before moving to the other company. Man, was I happy. I didn't think about money or anything. I was just happy because that was one step closer to SGT.

CHAPTER 7

My New Assignment to the 539th Transportation Bn

Five months after being assigned to the 539th Transportation Company, I was promoted to Specialist 5th Class, which is the first level of Sergeant. Per the promotion, I was assigned as Shift Leader of one of two cooking shifts. Shortly after being assigned to this position, I also took on the responsibilities of the Assistant Mess Sergeant. I must have been doing something right to be assigned all of these responsibilities with only 16 months in the United States Army. Let me explain how it all happened.

Shortly after I was assigned to my new company, there was another enemy rocket and mortar attack. Unfortunately, a soldier was killed when he left his bunker to go back and turn off the lights in his living quarters. The next day, I was able to unofficially look at the damage caused by the bombs. If I'm being real, I was just being nosy.

You see, I had never seen anything like this. I was really surprised as I tracked a piece of bomb shrapnel from the impact site. That piece of shrapnel traveled about twenty yards to the Mess Hall, went through a concrete wall, and continued through the back portion of a dining room chair. It then traveled approximately 25 to 40 feet and went through another concrete wall, where I lost the trail. But that was enough for me. I just thought that the death toll could have been a lot higher if the dining room had been occupied. Thank God it was not.

During that same period, there were a few things that happened that helped me get to the higher grade. Outside of doing a great job, there were a couple of unfortunate things that happened within the Mess Hall staff that created the Specialist E5 slots. There were eight E5s assigned to food service in my new Mess Hall. There was a Night Baker, two Shift Leaders, two Assistant Shift Leaders, and three other positions. I didn't waste my time trying to find out. I just went about my work.

Strangely, however, the E5s started disappearing. In addition, the Night Baker killed one of the other E5s and himself. Well, I guess I should explain. After the Night Baker got off work around 0700 hours, he filled a large commercial-size mixing bowl with iced beer. He started drinking that morning and continued drinking through the rest of the day until around midnight. Sometime during the early evening, some other soldiers, including an E5 Shift Leader, stopped by and drank beers with the Night Baker.

My friend told me that the Night Baker told the Shift Leader sometime during the evening that he was going to kill him before the night was done. According to my friend, who was lying in bed reading around midnight, he heard the Night Baker say that he was going to bed. According to my friend, the Shift Leader stated, and I quote, "I thought you were going to kill me before the night was done."

The Night Baker stated, "Oh yeah, that's right," then proceeded to his locker to get his weapon. Eight soldiers were sharing the cooks' tent. Our lockers were arranged in the middle of the tent with two rows of four lockers stacked back-to-back. I slept across from my friend at the end of the tent with one of the doors between us. At this point, my friend said that he could only hear the sounds caused by the weapon as the Night Baker removed it from the locker.

My friend said when he heard the Night Baker open his locker and lock and load his weapon, he sat up and listened more attentively. Then he heard a gunshot. That's when he jumped up and shook my feet as he ran out the door. As I woke up, I saw some people tussling at the opposite end of the tent. As I rose slightly, I heard a gunshot.

One person fell, and the other two started toward me in a hurry. My first thought was that we were under attack by the enemy. I just froze in place. Once they left the tent past me, I got up, got my weapon, and got dressed as fast as I could.

I saw the Night Baker lying on the floor in a pool of blood, and the Shift Leader lying on his bunk with part of his brain and lots of blood on the wall of the tent behind him. Still not knowing what really happened, I rushed into one of the adjacent tents and woke up those soldiers. Then I went back into my tent and continued into the other adjacent tent, where

I found my friend. Within a few minutes, the Military Police, the NCO who was in charge of the company for the night, and other leaders were on site. Those dead soldiers were listed as "Dead as a result of non-hostile action."

I really don't know what happened to the other E5s, except one of them became the Mess Sergeant. I replaced him as Shift Leader, and I took on additional responsibilities as the Assistant Mess Sergeant as well. For a short period, I also picked up rations for our Mess Hall from the ration distribution site (Ration Breakdown) using a large truck with a 40-foot flatbed. This job is where I heard a lot about the Black Market, which I knew nothing about at that time.

Civilians at the Ration Distribution Center and other soldiers were always asking me questions about where I worked, how far I traveled in the truck, and what type of supplies I hauled. If I had been dishonest or hadn't been so naïve, I would have been rich or in prison, probably the latter. Thank God for my naivety. One thing I remember is that, during the year I was assigned to that Mess Hall, we never got the furniture, air conditioners, and other items that we were told to expect. I am glad that I didn't get caught up in that Black Market mess.

CHAPTER 8

My Mess Hall Leadership Experience

I took pride in feeding the soldiers. I insisted that my cooks and KPs (Kitchen Police) be on time for work, and I ensured that their appearance and personal hygiene were above the normal expectations. Of course, food preparation as well as the appearance of the mess hall were at the top of my list. I made sure our food tasted great and had a pleasing presentation as well.

I also made sure our assigned soldiers were treated well when they came into our mess hall. Any soldier who came into our mess hall was treated with respect as well, which leads me to this experience. On this day, we were set up for lunch before serving time. The serving line and the food looked and smelled delicious. Suddenly, a soldier came in and asked if he could eat.

One of my cooks happened to be at the serving line. Lunchtime had not started yet, my cook stated. The soldier asked if he could eat again. The cook told the soldier that lunch would not start for a few minutes. He proceeded to tell the soldier that he could not eat in our mess hall because he was not assigned to our unit.

The soldier replied in a disturbed voice, "Sir, I am really hungry."

At that point, as I looked up, I noticed that the soldier was ragged and dirty, with his weapon slung over his shoulder. I quickly went to the serving line and interrupted the conversation. "No problem, man. Go there and get a tray. We can feed you now."

It was obvious that the soldier had been in combat for some time and was probably lost from his unit. I gave the soldier all the food he wanted. The soldier thanked me as he went to his seat. I counseled the cook on this type of situation.

First, the food belongs to the U.S. Army. Secondly, that soldier was obviously an American soldier who had been in a combat situation for what appeared to have been a long period of time, and probably got lost from his unit. Most importantly to me was that he looked very hungry. In this situation, it did not make any difference what unit the soldier was assigned to. FEED HIM. And we did. The soldier thanked me before, and after he finished his meal and continued on his way, I am sure, to find his unit.

On Thanksgiving of 1967, we prepared a meal for our troops that would have put the president's cooking staff to shame. (Smile.) The center table was about 20 feet long with three large golden-brown turkeys placed at equal distances apart, and all the Thanksgiving decorations we could put together. It was absolutely beautiful. The only thing missing was our families. But we made it through, and it was awesome! To this day, I still use some of the skills I learned during the preparation of that Thanksgiving Day feast.

45

CHAPTER 9

We Are Really All the Same

I learned a lot in Vietnam. It was my first experience working with and around people who lived outside of the United States. The people who worked as KPs in our mess hall and the cleaning ladies in our living quarters were all Vietnamese. There were other Vietnamese who worked in the Commander's offices and throughout the company area as well. I wondered for a long time why the ladies who worked in the Company Commander's and Battalion Commander's offices dressed differently from our KPs.

The administrative workers (ladies) wore long, silk-looking pants covered with a dress that seemed to be silk, split just below the waist down on both sides. They also always wore Vietnamese-style, wooden high-soled shoes. Our KPs always wore silk-looking pants without the dress, with flip-flops. I eventually learned that Vietnamese workers dressed just like we do on our jobs. All of us dress according to our job position; you know, business casual or blue jeans.

While working with the Vietnamese, I learned a little of their language. However, they seemed to learn more English than I did Vietnamese. We learned enough of each other's language, though, that we could communicate effectively. Working with them and communicating with them on many different subjects helped me to conclude that people are basically the same, no matter where in the world you live. The Vietnamese had, what is considered, different classes of people, just as the United States does. It's all driven by wealth or the lack thereof.

In my opinion, during this time, it seemed that Vietnam was way behind the United States in development. As I traveled around in Vietnam, I reflected on the information and pictures I remembered while studying geography in school. The upside-down, cone-shaped straw hat is what really caught my interest. In my geography book in high school, it seemed that most people in Vietnam wore that type of hat.

I also remembered reading and seeing pictures of people planting and harvesting fields of rice. Another thing I noticed was that our military presence provided lots of jobs for the Vietnamese economy, just as in the United States. We have civilian employees working on our military bases in America, and our bases in Vietnam provided jobs for the South Vietnamese. My life experience was greatly enhanced during my Vietnam tour. I learned a lot about the people and from the people.

Although I witnessed some frightening times during my tour in Vietnam, it was also very rewarding for me. As I mentioned earlier, because of my performance, I was advanced to Specialist E5 (Sergeant). I served in a leadership role for approximately eleven months out of the eighteen months I was in Vietnam.

Outside of work, most of my time was spent thinking about my family back in the United States. My fiancée, Sarah, occupied most of my thoughts. It was always exciting to read her letters. They made my fantasies of reuniting with Sarah as my wife feel as if I were already back home. I would receive at least one letter from her almost every day. There were times when I received several letters because of mail delays in the area. Fortunately, I later found a better way for us to communicate.

After we were married, I purchased two small tape recorders and sent one to Sarah. She and I communicated via tape during the rest of my tour. That really helped me tremendously because I hated writing, and my spelling was awful! I'm sure Sarah was elated because my voice was much better than my spelling, and she didn't have to interpret my meanings anymore. (Smile.)

Let me tell you a short story about my tape recorder. The friend I told you about earlier—you know, the pot smoker who loved to read—well, he became our night baker. One day after my shift, sometime in the early afternoon, I went to our living quarters in preparation for a nice shower and some rest. After about a minute in my area, I heard my wife's voice. I thought I was going crazy. Lo and behold, my friend was listening to one of the tapes from Sarah. I exploded.

What surprised me was that he thought it was ok. His statement was, "Man, you could let me listen to the tape. I don't have anybody to write me."

As I proceeded to remove the tape from his possession, I stated, "Not my problem, man. Sorry. Give me my tape! This is very personal, and I am not sharing my wife's messages with anyone."

CHAPTER 10

My Vacation Home / Dr. Martin Luther King Jr. Assassination

After my initial tour of one year in Vietnam, I received orders to serve in Europe. Because my fiancée was in college and would not graduate until 1969, I decided to take the option to extend six months in Vietnam, which allowed me to get a free 30-day leave at home. Also, after completing the additional six months, I would get my duty station choice in the United States.

As it turned out, I made an excellent choice. I never left the States again for the rest of my career. Keep reading, and you will understand how that happened.

During my vacation home, my fiancée, Sarah Harbin, and I were married during a beautiful, inexpensive, and surprising—to-some wedding. We really enjoyed each other's company for a few days before I returned to Vietnam to complete my six-month extension. Shortly after I left for Vietnam, Dr. Martin Luther King Jr. was shot and killed on April 4, 1968, in the city of Memphis, TN. Once I returned to Vietnam, I learned the news about the assassination and the riots that were taking place because of Dr. King's assassination.

While Black Americans were fighting alongside White Americans in Vietnam for the freedom of the South Vietnamese, it appeared that Black Americans were fighting White Americans back in America because of the assassination of Dr. King. I knew then, and I have found out since, that every person who was rioting was not doing so because of Dr. King. A lot of those people used that situation as an opportunity for self-preservation. People were burning businesses as a cover to take anything they could get their hands on. It is unfortunate, however, that this is what it takes to force politicians to do what it takes for the preservation of all American citizens and not just some.

CHAPTER 11

My First Stateside Assignment

In September of 1968, I was assigned to Fort Jackson, South Carolina. My decision to extend my Vietnam tour proved to be extremely beneficial to me and my eventual family. During my next 22 years of U.S. Army service, I never left the United States again. Keep reading. I will explain.

I was assigned as the Shift Leader of one of the two cook shifts in the Headquarters Company Consolidated Mess Hall (cafeteria) of the Infantry Training Brigade. This was an extremely large Mess Hall which was designed to feed, I believe, up to 500 soldiers per meal. I have got to tell you that this Mess Hall had an extremely well-equipped kitchen.

Of course, having been in the U.S. Army for only two years, this was my first experience working in a Mess Hall of this magnitude. Yes, I was a cooking-shift leader in Vietnam; however, the Vietnam Mess Hall didn't compare to this one.

Despite this, I adjusted very well to both my new position and the totally different surroundings.

Although the Brigade Headquarters Company only had approximately 150 soldiers assigned, my shift was responsible for feeding as many as 400 soldiers per meal from other companies within the Brigade. My team consisted of as many as ten soldiers. My immediate supervisor was a Sergeant First Class (SFC/E7), who was Black. Remember, I point out race only because I want to make a point about discrimination.

This assignment was where I learned that discrimination has many faces and reasons. The first discrimination I experienced by another Black person was within days after I started work in my new position. Check this out: my Mess Sergeant asked to borrow my car. I declined but offered to take him where he needed to go. He got very upset and stated in a very hostile voice, "That's all right, dipshit. I will remember this."

I really didn't know what he meant at that moment, but I sure found out as time went on. It seemed that he found any reason to screw with me for any little thing. For example, he was filling a box he had placed by the back door of the kitchen with food products from our food supply room. I had been in and out of the supply room as I did many times when my shift was on duty.

As I was leaving the supply room on one trip, the Mess Sergeant told me to take the box and put it in his car. Knowing this was against Army policy, of course, I declined but offered to have one of the KPs (trainee soldiers who worked in the Mess Hall as part of their scheduled detail) do it. After I declined, once again the Mess Sergeant exploded, stating, "That's all right, dipshit" (that was his pet word). "I'll do it myself."

I just motioned with my shoulders, "ok," and went about my work. He was obviously pissed. Did I care? No. However, I remained respectful of his rank. I didn't have to wait long before the Mess Sergeant acted on his threat. Keep reading.

I ran a very tight shift. I demanded excellence of my cooks, which included being on time all the time. For example, one of my cooks was so late coming to work that I had to assign his work, for the lunch meal, to another cook. The assignment included cooking meat on a commercial griddle. When the late cook finally came to work, I had him clean the griddle since he was assigned to cook the meat for that meal. Instead of doing what I told him to do, the cook complained to the Mess Sergeant.

I had no problem with him complaining. However, I did have a problem with the Mess Sergeant taking the cook's side instead of supporting my decision as the supervisor, which would have been the right thing to do. To no avail, I explained my reasoning to the Mess Sergeant, who seemed to take pleasure in going against me with my subordinate. I must point out that the same soldier was really a slow learner and a poor performer, and the Mess Sergeant knew that.

On another occasion, I assigned the same cook to prepare Onion Soup, which was a very simple task. Each cook is required to always use the U.S. Army Cookbook, whether he knows the recipe or not. Now, it wasn't hard for me to notice this guy making a mess of this food item because I always paid attention to my men, not only with food preparation, but everything they did while on duty. This cook was at the forefront of my attention.

As I moved through the kitchen, checking out everything, as I always did, I watched him as he filled the huge pot with enough water to feed perhaps two or three times as many troops as we had to feed. Then I noticed that he did not have the soup recipe. As I approached to correct his actions, I noticed him starting to peel only five large onions.

As he started to drop onions in the full pot of water, I asked him to explain how he was going to make the soup. This poor guy proudly told me that he was making Onion Soup, but he never looked in his cookbook. This cook never explained what ingredients he would use to make the soup. Therefore, I gave him directions and watched him as he made the soup using the recipe from the cookbook. Man, was he pissed, but no problem. I didn't mind him being pissed. I was glad that I caught him before he ruined that portion of the meal.

The second retaliation was during a routine inspection conducted at the Mess Hall by higher headquarters. These types of inspections were not unusual. They were designed so that we would maintain high-quality Mess Halls for our soldiers. Anyway, the inspection was during the other shift's tour. The shift leader on duty was always responsible for preparing for any inspection that was conducted during their tour of duty.

To help the shift leader prepare for the inspection, I offered to bring my shift in early to help them prepare the noon meal so that his people could concentrate on the inspection. My counterpart declined my help, and I had no idea why. Since he was the Mess Sergeant's fair-haired child, I guess he thought the Mess Sergeant would take care of him just as he had done many times in the past. I'll explain later. Keep reading.

They failed the inspection, and the Mess Sergeant went absolutely berserk. He had a meeting with all the cooks from both shifts, and it was not good. It seemed that he used every curse word I had ever heard. I hated that they failed the inspection, but I couldn't do anything about it because, as I stated, the shift that's on duty during an inspection has full responsibility for preparing for the inspection.

It seemed, however, that the Mess Sergeant directed all his anger toward me. As he dismissed us, he proceeded to single me out. He got in my face, or rather my chest (you see, he was kind of short (Smile)), used several choice words, and proceeded to tell me, "And you, dipshit, you walk around like you think you are too cute to do any work. I want to see you with a broom and mop in your hand doing some damn work."

Man, was I pissed. He unfairly called me out in front of my men as well as the soldiers assigned to the other shift. I remember standing there with the Mess Sergeant screaming at me, hearing this voice in my head (the devil) saying, "Hit this SOB. You can take him."

Then I heard this other voice saying, "Think about your family."

In later years, I realized that was an intervention from God. Instead of hitting the Mess Sergeant like I really wanted to, I just turned and quickly walked to the door leading into the kitchen. When I kicked the kitchen door to open it, the door vent went flying into the kitchen. I quickly walked through the kitchen and went directly into the latrine (restroom), where I literally cried like a baby.

I cried for a few minutes, getting all the anger and frustration out of my system. Shortly afterward, I gathered my men and had a brief meeting. I told them that I wanted all of us to take a moment and put everything that happened out of our minds. Then I wanted us to go about our work and get ready to feed our soldiers their dinner meal. My men had no problem following my instructions.

As I mentioned earlier, I was very demanding. First, be on time for work; no excuses. If you don't feel good, go on sick call. Secondly, you must be well-groomed. Your personal hygiene must be top-notch, and your uniform must be clean. I would check my cooks at the beginning of each shift. A dirty uniform is a NO-NO.

I would always send anyone who violated my instructions back to their room to fix the problem and report to me for re-inspection. I accepted no excuses for cleanliness. All cooks received free laundry, and I insisted that they use that service. After all, we were preparing food for hundreds of soldiers who could see us at some point during the day, plus your job puts you around food your entire shift. Therefore, I held my cooks to a very high standard, especially with their personal hygiene.

I assigned each of my cooks to prepare a segment of the meal. Every item on the menu was assigned to a designated cook. This included the meats, vegetables, starch items, desserts, salads, soups, and drinks. I also assigned a cook to prepare garnish for the serving line. I was not a black-and-white supervisor as it relates to our Mess Hall regulations and manual of operations. However, I read these documents and made sure I, as well as my team members, understood them.

There was a sergeant from within the units we served, assigned as headcount for each meal. Part of their responsibility was to keep a count of the number of people that we fed during that meal. This included those people who were required to pay for their meal because they received Separate Rations (funding to eat outside of the Mess Hall). Each headcount had to report to me, as shift leader, for a briefing. I made sure they understood that each paying customer must pay for the meal prior to consumption. I quickly learned that the Mess Sergeant had allowed some of his friends to eat without paying for the meal.

I had a couple of incidents where one of those friends tried to defy the rules. In each case, the headcount followed my directions and reported the incident to me prior to the Noncommissioned Officer being served. In one of the cases, the Noncommissioned Officer (the Mess Sergeant's friend) tried to "pull rank on me." The conversation went something like this:

Me: "Good morning, Sergeant First Class. Can I get you to pay for your meal?"

SFC: "No, I will pay after I finish eating."

Me: "I'm sorry, Sergeant First Class, our regulation (I quoted the regulation number)... Boy, did that piss him off." (Smile.)

SFC: "Don't be quoting regulations to me, Specialist," he stated.

He was hostile but proceeded to pay for his meal.

Sometime later during my tenure, one of the Mess Sergeant's friends asked the Mess Sergeant, "Does that specialist ever do any work? His uniform is always clean." Instead of the Mess Sergeant seeing this as a positive thing, I was told by someone who heard the conversation that he tried to turn the whole thing into a negative, saying something like, "That dipshit ain't worth a damn."

No problem. I just kept doing my job. To this day, I take pride in not letting hearsay get into my head. I never let anything or anybody keep me from doing the best job I could possibly do, no matter where I was assigned.

This next segment of my story is one of the times that I felt that God was guiding my life. Keep reading.

Sometime later during my first year, I had two negative events while feeding the infantry trainees in the field. On one occasion, the Deputy Base Commander visited my station while I was on duty to investigate why several trainees got sick after the noon meal. He explained to me why they were there and proceeded to have his medical staff conduct a full inspection of the Mess Hall, food, and cooks. It was determined that the training area had been sprayed with insecticide earlier that morning, which caused the problem. They were supposed to allow 24 hours before allowing troops into the sprayed area. I was curious but never worried because I always made sure my cooks were clean and medically fit to work in the kitchen and prepare food.

After the meals were prepared and we were ready to feed the troops, I always used this time to take a break and give my cooks the opportunity to eat their meal and relax for a few minutes before feeding our troops. I also used this time to just communicate with my cooks. No business; we just talked.

During this period, a while later in my first year, the same general who visited a training company as we fed them lunch visited our Mess Hall again, but for different reasons. The general and his aide-de-camp (personal assistant) entered the Mess Hall through the side door directly next to the serving line. This was the door that was intended to be used for the troops to enter the dining room and go directly to the serving line. However, because there was no cover for the soldiers outside, the leadership decided to have the troops enter through the front door, which allowed them to line up along the wall inside the Mess Hall.

Once the general and his aide were inside the Mess Hall, my senior cook informed me that someone was standing by the serving line. As I approached the two soldiers, I realized who it was and proceeded to take the necessary action.

"Sir, Specialist Wilson reporting. Welcome to our Mess Hall."

The general replied, "Hello, Specialist Wilson. I'm General Coleman, Deputy Commander of Fort Jackson. I visited the troops you fed in the field. I was very impressed with the layout. So, I decided to come in and see if the service was as good and the food as well prepared here in garrison as in the field. Tell me what you have here."

At that point, I thanked the general and proceeded to explain each item on the serving line from the first item to the drinks. As I introduced a specific item, the general questioned why that dish was not a part of the field chow. My response was, "Sir, this item was left over from our noon meal, and I decided to serve it to the troops as extra. Leftovers must be served during the next meal in-house. They will never be served as part of the food we serve in the field."

As I completed my briefing of the serving line foods, the general stated, "Excellent, Specialist Wilson. Let's go to the office and see what's going on in there."

I didn't get nervous (YEAH RIGHT, smile), but my thought was, "I don't know jack shit about that office stuff." But, to my surprise, I was awesome. (Smile.) I actually surprised myself. I really didn't realize that I knew all that I did about Mess Hall operations. I felt great about that.

As I continue, I appeal to you to please understand that I did not intentionally dictate the outcome of what happened because of the general's visit. However, the action that the general took was a blessing for our Mess Hall. Keep reading.

Once we were in the office, the general asked me several questions pertaining to the office administration and the Mess Hall as a whole. After about 10 or 15 minutes in the office, the general stated, "Everything is great, Specialist Wilson. However, I have one concern. Why didn't you have a container for the troops in the field to drink their milk from?"

Man, was I surprised. Although it never occurred to me until the general asked the question, I knew exactly what happened. After a brief pause, I explained, "Sir, ration breakdown (the food distribution section) usually sends our milk in half-pint containers from which the troops drink the milk. This time, the milk for today was sent in two-gallon containers, which are meant to be dispensed from the milk machines. I'm sorry, sir. I forgot to send cups for the troops to drink from. That was my fault."

The general stated, "No, it wasn't your fault, Specialist Wilson. Where is your Mess Sergeant?"

I stated, "Sir, he leaves every day around noon. I think he has a part-time job in the city."

The general: "Where is the Assistant Mess Sergeant?"

I stated, "Sir, when I am on duty, I see him leave just before we serve the dinner meal."

As I escorted the general and his aide to the front exit of the Mess Hall, he stopped at the headcount station. When the sergeant did not stand in respect for a senior commissioned officer, I ordered him to the position of attention. The general turned to face me and stated, "Specialist Wilson, I want you to have your Company Commander report to me in my office at 7:30 tomorrow morning."

I said, "Yes, sir," and rendered a departure salute. Just as I completed my salute and turned around, my First Sergeant (the senior enlisted soldier in the company) approached the general and reported.

In just a few minutes after the general left, the Company Commander, First Sergeant, Company's Executive Officer, the Mess Sergeant, and Assistant Mess Sergeant met in the Mess Hall office. They drilled the hell out of me. They wanted to know everything the general said and did, and why he came to our Mess Hall.

Of course, I was obliged to tell them everything. Make no bones about it, I made sure the Company Commander got the message to report to the general the next morning. What was really great about the whole thing was that my staff continued with our responsibilities. They fed the troops, began prepping for the next meal, and started cleaning for the night. I thanked them collectively for their loyalty and let them know that I was extremely proud of what they did in my absence. I also made sure I complimented my specialist, who was in charge during my absence, for stepping up while I was busy with the general.

The next morning, the Company Commander returned from his meeting with the general. The Company Commander approached me as I checked the roast beef that was being cooked for the noon meal. I could tell he was not happy. Twice, he passed on to me a message from the general. Each time he stated what the general said, he spoke so low that I could not understand what he was saying.

It was obvious he did not have a good meeting. I told my Company Commander for the third time that I didn't understand what he was saying. After he managed to clear his voice, he told me that the general told him to tell me that he was very impressed with talking with me during his visit.

I simply responded, "Thank you, sir." I smiled and continued with my work. Of course, I felt extremely excited about that. I don't know what happened to the Company Commander, the Company Executive Officer, and the First Sergeant. However, I don't remember ever seeing any of them again. I was told that the Mess Sergeant was shipped out to Vietnam, and the Assistant Mess Sergeant retired. Anyway, I never saw either of them again.

I didn't know much about the company's leadership because I never interacted much with them. However, in my opinion, the Mess Sergeant was not a good person. He broke many of the Army's regulations, and he was not a people person. He didn't care about anybody who didn't kiss his ass. So, I believe moving him helped our Mess Hall tremendously. God works in mysterious ways.

Within a very short period, a temporary Mess Sergeant was assigned. He introduced himself to our kitchen staff and went about his work. He never got involved with my staff or me. I really don't know if he needed to get involved with the other shift or not. I do know that our previous Mess Sergeant had spoiled my counterpart. He could do no wrong in the eyes of the previous Mess Sergeant.

One very noticeable thing was, it seemed my counterpart would always go out of town during his shift's four-day off period (from Friday noon to Monday noon). The problem, as I saw it, was that he was always substantially late returning to work, which was, seemingly, no problem for the Mess Sergeant. I didn't let that concern me. Regardless of what was going on with the other shift, I just made sure I took care of my responsibilities.

CHAPTER 12

The New Mess Sergeant

Approximately a month or so after the temporary Mess Sergeant was assigned, our new Mess Sergeant, who happened to be white, reported to the position. Just as a reminder, as I mentioned earlier, I point out race simply to clarify details in my story. Keep reading.

The new Mess Sergeant shadowed the temp Mess Sergeant for two weeks. He never got involved with anyone on my staff except to speak. I noticed, however, that when he was visible, he inconspicuously watched what was going on as he walked throughout the Mess Hall, with or without the temporary Mess Sergeant. After two weeks, he took charge of the Mess Hall.

The very first time he said anything to me, other than a greeting, he came directly to me and said, "Specialist Wilson, I'm going to see to it that you get promoted. I like the way you function."

All I could do was smile and say, "Thank you, Mess Sergeant." We always addressed the Mess Sergeants by their position, not their rank.

One of the things I have learned from experiences I encountered during my life is that we should never judge anybody by their race or the color of their skin. Unlike our first Mess Sergeant, our new Mess Sergeant was a blessing to me, my staff, and the Mess Hall. He saw no race. He did what great leaders should always do, which is take care of their people, no matter who they are. You cannot be an effective leader if you are a bigot. In my opinion, there should not be a place in leadership for a person who hates another for any reason.

Remember my counterpart, the other shift leader? Well, he tried his usual call-in-late tactics with the new Mess Sergeant. Word was, he was told that he would be at work on time and that excuses would not be accepted. He tried it again to no avail. The Mess Sergeant did what the old Mess Sergeant should have done. I was told that the new Mess Sergeant pressed charges against the shift leader. I can tell you he was punished because I noticed that the rank insignia on his uniform had been reduced by one grade.

My Surprise Promotion

One Sunday afternoon, when I was just beginning my second year, my shift was on duty. Business was slow, and we were serving cold cuts, soup, and salads, so I had lots of time on my hands. I decided to walk up to the Consolidated Mess Hall next door. I had never visited either of the other two Mess Halls, so this was a great opportunity for me to meet my counterpart. It ended up being a Godsend for me. Keep reading, and you will understand why.

My counterpart at the Mess Hall I visited met me at the back door. I introduced myself, and he did as well. We talked for a moment, during which time I noticed his rank: Specialist 6 (SPC/6), which was the grade for our position, a grade higher than my present grade. The conversation went something like this:

"How long have you been an SPC/6?" I asked.

The Specialist stated, "They just promoted me this month. Yeah, man, they promoted me, hoping I would stay in this man's Army. I will get out in a few weeks. I am not staying in this fucking Army!!!"

As he went on and on about his promotion, I continued to record everything he said in my memory bank. I left after about thirty minutes and returned to my Mess Hall. I couldn't wait for Monday morning so I could tell my Mess Sergeant what the Specialist had told me. The Mess Sergeant had not said anything else to me about promotions since he told me he was going to get me promoted. You see, I trusted what he told me. He was that kind of leader.

On Monday, after I found the appropriate moment, I laid my lips on the Mess Sergeant's desk. (Smile.) That is to say, I told him everything the Shift Leader next door told me. The Mess Sergeant got up from his seat and went straight to Brigade Headquarters. I don't know everything he told the Battalion Sergeant Major, but sometime late Monday morning, I was promoted to pay grade E6.

What I was told was that my Mess Sergeant went into Brigade Headquarters, raising holy hell (in a nice way, smile) about the promotion next door. He told them that he wanted me promoted and why. They responded that they didn't have any more allocations for a SPC/6 Food Service Shift Leader.

The word was that the Mess Sergeant asked if they had any unused infantry allocations.

It so happened that they had allocations for Infantry Staff Sergeants (E6). My Mess Sergeant suggested that they promote me using an Infantry Staff Sergeant allocation and amend the promotion when the food service allocations came in. That's what they did.

Check this out: during the ceremony, they pinned the chevron on the left side of my shirt. When I got home, I walked around for a few minutes to see if my wife would notice. Poor baby, all she said was, "Why do you still have on your uniform? Aren't you going to take a shower?"

I said, "Baby, do you notice anything different about me, about my uniform?" Of course, she said no.

I said, "Baby, I got a raise today." With a big smile on my face, I stated, "I am now a SPC/6 shift leader."

Nothing clicked in her mind except the raise part. Lord knows we needed the money. Sarah knew nothing about the military or its rank structure, but she was great at managing our finances. Her accounting skills are the reason we got through those tough early years of my career, and she continues the same to this day. My wife always supported me in whatever I did and wherever I went throughout my career, and I did the same for her during her career with the federal government.

I let the Mess Sergeant know how much I appreciated what he did for me. He told me that I did everything, that I earned the promotion, and that it was his responsibility to fight for me or any of his soldiers who did a great job. He went on to say that if I kept performing the way I had been performing, many great things would come to me.

Let me tell you this: the promotion didn't change me. I continued to do what I do. I made sure my cooks always prepared exceptional food for the troops. For example, usually, soldiers couldn't wait for payday, so they could go out into the city to eat. They wanted to eat anywhere but the mess hall. When my shift was on duty during a pay period, our mealtimes were like any other day.

When the soldiers found out that my shift was on duty, most of them didn't go out to eat on payday. They saved their money and came to our Mess Hall to eat because they knew my cooks served great food. Our meals always embodied the word *appetizing*.

To this day, I am convinced that if the black Mess Sergeant had not been shipped out, I would not have been the recipient of this promotion or any of the advancements I received during the next few months. He disliked me because of such childish things as my not letting him use my car and not allowing one of his friends to continue consuming meals without paying.

First, a manager is not supposed to fraternize with their subordinates. Also, soldiers with dependents receive extra funding for rent and meals. Therefore, when they eat in our Mess Halls, they must pay a cost that is set by the United States Government for each meal. What the Mess Sergeant knew but failed to properly enforce was that the soldiers eating in a specified section of the dining room were on separate rations.

I try not to ever bring politics into my work. That is why I always read and understood the regulations pertaining to my responsibilities. In this case, I was obligated to enforce all rules pertaining to Mess Hall operations.

Anyway, I continued to do what I did, and it paid off. In a very short period after I was advanced in grade, I was selected to pilot a new program designed to help infantry instructors and trainees get a good start to the day during the winter months. My responsibility was to carry morning food and refreshments, coffee, juices, and other items to the campsite of the chosen infantry company. I would set up a partial field kitchen equipped with a field range, equipment, and condiments for coffee and the other items that would serve a company-sized unit of infantry soldiers.

I also kept records on travel time and mileage to and from the site, set-up time, breakdown time, and how much time was needed to serve the number of soldiers in camp. The company was responsible for letting me know how many troops they had on site and what time the "first call" was (the time the soldiers were awakened in the morning). Most mornings, I had to be on site by 4:00 a.m. Man, I hated that. But I did what I had to do to get the job done.

I was successful in compiling the necessary information that the command needed to decide whether to implement the program or not. I recommended the implementation of the program because I felt the initiative would help the troops get a great start to the day, at least for two out of five days of combat training.

CHAPTER 13

My Assignment to The General's Staff

I never got a chance to see if the program was implemented. Unbeknownst to me, the General, who had made a random visit to my Mess Hall earlier that year, already had his eyes on me. Let me explain. General Coleman was now the Base Commander of Fort Jackson, South Carolina. For those of you who don't know, the Base Commander position is comparable to the mayor of a city. Well, he contacted my Brigade Commander and asked him to check with me to see if I would work for him. The assignment would be temporary while his Enlisted Aide (Chef) was on leave.

I was told by my Mess Sergeant, with a smile on his face, to report to the Brigade Sergeant Major right away. Now, I knew I had not done anything wrong, but that did not stop me from wondering what this man in such a high position wanted with me, a peon.

During my meeting with the Sergeant Major, he informed me that the Base Commander wanted someone who could cook for him while his assigned Enlisted Aide went on a 21-day leave. He went on to say, and I quote, *"You are the only one that I can trust to do that. We can't order you to work in the position. You would have to agree. What do you say? Would you like to do this?"*

Of course, I agreed. I never refused an assignment, especially when a superior thought enough of my work to request my services. This was an assignment I would not have dreamed of in a million years—Specialist Wilson, the Base Commander's personal cook. Go figure. He must have seen something in me that he liked when he visited my Mess Hall. I am confident he spoke with my superiors as well.

I reported to the Major General's house one morning in November 1969. I was introduced to his wife and three children. I received a briefing from the General's Enlisted Aide, who I would be replacing for a few weeks. The General's family consisted of a daughter, Mary, who was thirteen years old; a high school son they called Bill; and an older daughter, Nancy.

We were only supposed to cook for the Base Commander; however, on occasion, I gladly did things for the other members of the family when needed. They did not fit the typical perceptions you hear about people in high places. They were a wonderful family. I'll share more interesting information about them later.

The Birth of Our First Child

First, I must share this extremely exciting event in our lives with you. Sarah and I were blessed to bring our first child,

Dimey, into the world. Dimey is pronounced "Diame'me." I chose the name by closing my eyes and hitting the keyboard five times, which gave me "Diame." However, because people were pronouncing her name as "Demi," like Demi Moore, I changed the spelling to Dimey (though some still mispronounce her name).

What really got me was, during a lunch break with my wife, Sarah, in a park in downtown Columbia, SC. A retired English schoolteacher sat next to us. As Dimey played on the grass, the lady asked what her name was. I answered, pronouncing our daughter's name the way we wanted it. The retired English teacher confirmed the name by spelling it according to my pronunciation. She spelled it exactly the way I spelled it initially. I was so upset that I considered changing the spelling back to my original spelling, but I didn't. Life goes on.

Anyway, our oldest will soon be 56 years old. Being a daughter of military parents, Dimey probably holds the record for the number of schools she attended during high school. (Smile.) Between Amarillo, TX, Knoxville and Nashville, TN, she attended three different schools between the ninth and twelfth grades. No problem, this did not hinder our very smart daughter from doing well in school.

Dimey graduated from John Overton High School in Nashville, where she was a very successful band student. She surprised us by playing a solo during one of the band's last concerts of that year. Sarah and I were extremely proud of Dimey, not only for the solo she played so well, but also because, on her own, Dimey applied for and received a band scholarship to college at Memphis State University, now (The University of Memphis).

Dimey and her husband, Jeffrey Sharpe, gave us two wonderful and lovely grandchildren. Our grandchildren, Brittney and Jeffrey Jr. (Tyrone), are doing extremely well as adults in their respective careers. Brittney has worked for a well-known communication company for several years. She also owns a skin care company named *Butterfye Evolution,* where she produces an organic body butter. Tyrone has worked in Hotel Security in the Nashville area. He is now a School Resource Officer and a volunteer Deputy Sheriff.

Back to my assignment on the General's Staff

About a week or two after I started the assignment, I received orders to return to Vietnam. The General usually asked me how things were going with me and my family when he was home for lunch. On this particular day, when he asked how things were, I told him about the return-to-Vietnam assignment.

The conversation went something like this:

The General: "How are things going with you, Henry?"

Me: "Not too good, sir."

The General: "What's going on?"

Me: "I just got orders reassigning me to Vietnam. Sir, I don't have a problem with the assignment; it's just that the timing is messed up. We just had our first child. If I leave now, she will be over a year old when I return home. There is no way she will know who I am when I get back home."

The General paused for a moment and then said, *"I'll look into it and get back to you."* I said, "Thank you, sir," as he departed to go back to his office. I couldn't wait to tell my wife when I got home. We were very hopeful, but our hope faded when the General didn't say anything for almost two weeks.

We began to face the fact that I was going on the assignment, until one morning, after he finished breakfast and started to leave for work. He stated, *"I'll see you later, Henry."*

"Ok, sir," I said.

The General paused for a moment and stated, "Oh yes, I'm still working on your problem. *As a matter of fact, I will call you shortly."*

"Ok, sir, thank you," I responded.

At that point, my entire body was filled with hope and excitement because of the anticipation of what I hoped was to come. However, I refused to tell Sarah until things were settled.

As promised, the General called me after he returned to his office and checked out a few things. The news was excellent. *"Henry, this is General Coleman."*

"Yes, sir," I responded.

"The only way I can get you off the assignment to Vietnam is if you continue to work for me, but it must be voluntary. Do you want to continue to work for me?"

I wanted to say, *You damned right!!!* But I controlled my emotions and kept it professional. "Yes, sir!"

"Ok, it's done," the General responded. *"Your orders are cancelled, and new orders are being prepared for this job."*

"Thank you, sir. I really appreciate this."

With tears in my eyes, all I could think was, *"God is really working on me."*

You see, this position came about because the General was advanced in grade to Major General and became the Base Commander. In his position of Commanding General of Fort Jackson, he was authorized two Enlisted Aides. All this took place soon after I was assigned to the General's staff. Once again, I recognized that God had intervened in my life. You better believe that I let Him know that I appreciated what He did for my family and me. Of course, I called Sarah right away and gave her the good news.

This turned out to be a really great move for my family and me. My duty station was—get this—the General's house. The Base Commander, Deputy Commander, and the General's Chief of Staff all lived in houses located in a secluded, restricted area next to a small lake. During the short time I worked there, I got to know the family very well. The thirteen-year-old, I guess you deduced that she was the youngest of three children, required a lot of my time, which I did not mind at all. Since the General was away a lot, as well as his wife having lots of community responsibilities, I was the go-to person when a parent's attention was required.

I acted on calls from school for both younger children, but primarily their baby daughter, and I provided transportation using the family car when needed. I picked the younger daughter up from school and took her to bowling lessons on Saturdays. I got to tell you! It was awkward when I drove the General's car. Since the family car was marked with the General's insignia, soldiers were required to render a salute as we approached them. But if you know me, you know I handled it well. I never took advantage of this privilege. I would just wave to those who rendered a salute. In addition, my route was always directly to and from the primary destination.

Their baby daughter was a sweet young lady who required lots of my attention, which I didn't mind. At one point, the mother came to me as I prepared the General's uniform with a concern that most parents face, especially when the parents are out of the home a lot.

The conversation went something like this:

The First Lady: *"Henry, can I talk to you for a minute?"*

"Yes, ma'am, of course," I responded.

"I have concerns with my baby daughter. I don't know what I'm going to do with her. I am having concerns getting her to complete her daily responsibilities."

I thought under my breath, *here we go again.* However, if you converse with me or ask for my opinion, I will be completely honest with you. If you ask me a question, I'm going to tell you what I think. Although I was kind of surprised that she came to me with her concern, I proceeded with an answer.

"Ma'am," I responded, "I think it is a good idea that we tell our children what we want them to do and add a completion date and time when appropriate. What is important is that we make sure we follow up with our demands as soon as possible. Of course, it is always important to give our children continual love. But be sure to follow up by checking your demands in a timely manner. If the task is completed, praise them. If they did not complete the task, I would have them complete it right away. If they need help, help them."

The conversation continued for a while. Then the mother thanked me and, with a smile, went about her day. These were two intelligent parents. Why did she ask me that question? I think she may have been testing me. Who knows?

Anyway, back to my duties, which also included, of course, mostly cooking breakfast and lunch for the General. On occasion, the wife would ask for a casserole to be prepared and set aside for their dinner meal. We also prepared food for social functions that hosted dignitaries from all over. The Senior Enlisted Aide (Chef) prepared most of the food for these functions, while both of us served food and drinks during the events and, of course, cleaned up afterward. Sometimes the General would invite us to sit and commune with him.

He would ask a lot of leading questions as we sat and talked. This was his way of learning about our personal lives, what was happening in the community, and the overall climate on base. Surprisingly, we had very good and informative conversations, mostly about our military lives and our families. It was really great. I knew his intent, and I was obliged.

On another occasion, after he had finished his lunch, he started a conversation with me that went something like this:

The General: "Henry, my Commanders (Senior Officers who hold command positions throughout the base) tell me that the morale among the soldiers on base is very high. What do you say about that?"

It was like he knew I was going to shoot straight from the hip, and I did. I took a deep breath, thought for a few seconds, and then responded:

Me: "Sir, the only way you can gauge the morale of the soldiers on this base—or any base—is to spend time with them in their environment. A lot of your Commanders are going to tell you what they think you want to hear, not necessarily what's factual. The worst part is that a few of them have not done their jobs, and in some cases, they really don't know the in-depth responsibilities that their job entails. Therefore, they unwisely deceive you in hopes that they can fix the problem before it can be checked out.

"Sir, history shows that Senior Commissioned Officers don't normally follow up because they depend on—and they should—the honesty of their subordinate officers, both Commissioned and Noncommissioned. Commanders in senior command positions must weed out those they cannot trust. On occasion, it's a good practice to follow up with subordinate officers and Noncommissioned Officers to ensure they are in compliance with the mission.

"In the case at hand—morale—I recommend that you visit some of our units and talk with the soldiers at their training sites or their workplaces. I think this is the best way to gauge the soldiers' feelings about their service in the Army and about being away from their families."

When I realized I must be talking too much, I apologized to the General, to which he responded, "No apology necessary, Henry. I appreciate your input. Thank you."

Then he departed to return to his office. Shortly after the conversation, I realized that the General already possessed those qualities we talked about. After all, he found me by being directly involved with the soldiers when he was the Deputy Base Commanding General (DBCG).

Not long after our conversation, the Deputy Base Commanding General (DBCG), who was responsible for all training on base, started spending lots of time at the training sites of the units on base.

One of the first funny things that happened was when a Company Commander walked into his office early one morning and found the DBCG sitting at his desk. I must point out that this is hearsay; however, I believe it. I have witnessed the actions of supervisors when they are expecting a visit from their boss. Therefore, I am not surprised that the Company Commander supposedly almost jumped out of his skin and stepped into the trash can as he reported to the general. Man, just thinking about that makes me laugh. The DBCG spent countless hours with the training units.

Ok, back to my experience at work with the Base Commander and his family. Early on during my assignment with the general, I felt comfortable that he was a fair man. He did not care for people to suck up to him. I experienced one incident that made the general angry.

The Base Commander asked me to take his daughter's TV to the repair shop, which was on base and operated by a civilian contractor. Of course, when I dropped the TV off, I put it in the General's name and in care of me. In a week or so, the General gave me a check to pay for the TV. When the General returned home for dinner, I gave the check back to him with the following message: "Sir, the contractor gave me the check back and stated, 'Tell the General that since he is the Base Commander, there is no charge.'"

Obviously, the General became very upset. He handed the check back to me, saying, "Henry, take this check back to that man and tell him that if he pulls another trick like this, I will run him off this base."

I responded, "Yes, sir."

Man, I could hardly wait for the next day because I knew that man would have, or had already, cheated some lower-ranking soldier out of that same amount of money or more to cover his losses. The next day, I took the check back to the contractor. I took extreme pleasure in looking him in the eyes and repeating exactly what the General said.

This is what I said: "The General told me to give this check back to you and to tell you that if you pull another trick like this, he will run you off this base."

I could see his disgust when his face changed as he turned and went to the back of the store. Man, did I feel good as I turned, smiled, and left the store. As I mentioned earlier, we did lots of social functions with the school-trained Enlisted Aide doing most of the main course preparation. However, I held my own when it came to traditional food preparation.

Quick story: Once I was asked to go to the DBCG quarters and assist their Enlisted Aide in setting up for an Officer Wives' function hosted by the Second Lady. The guy had a layout like nothing I had seen before. It was super. But, in my professional opinion, there was one mistake he made that I tried to correct. He was barbecuing chicken that he had soaked in a thin marinade.

He had started the charcoal that was burning bright red in an open grill. When I saw that he was placing the chicken on the grill, I tried to stop him. I recommended that he wait until the charcoal burned down more before placing the chicken on the grill. Otherwise, he would have to stay with the chicken because the fire was too hot, and the sauce would cause the chicken to burn right away. He didn't listen and turned and went back into the kitchen, saying, "It'll be all right."

Thinking that the Chef had something else in mind to keep the chicken from burning, I left too. As a matter of fact, I went back to the Base Commander's quarters because things were pretty much ready except for the chicken. As time went on, I regretted not staying and assisting him with the chicken entrée. I may have learned something. To this day, I wonder how he handled it because I was sure that the chicken would burn if not turned repeatedly.

Although I did pretty much anything the family asked of me or that I thought needed to be done, I was mainly responsible for the General's uniform and his food. He wore the Class A Uniform almost all the time. The ribbons, medals, awards, and insignias worn on the General's uniforms were somewhat different from how enlisted soldiers wore their uniforms. I did research on preparing his uniform, and since generals had a choice on how they wore these items on some parts of the uniform, I simply used the uniform that he wore as a guide.

I also prepared his breakfast and lunch meals when he was not traveling. As part of my duties, although rarely, I also prepared his dinner meals. However, once I prepared his dinner meal, this situation stuck in my mind vividly. I prepared a sirloin steak cooked on a charcoal grill, a baked potato, and a garden salad with tea. I arranged the food on a table I set up on the patio.

As the General sat and started to eat, I proceeded into the kitchen. A minute or so after I got back into the kitchen, his wife came in and told me, "Henry, the General wants to see you." My heart went to my stomach as I said under my breath, AAHHH SSSHIT, what's wrong?

"Yes, sir, what can I do for you?"

"Henry, this is the best steak I have ever tasted. How did you cook this?"

The Spaghetti Sauce Story

I got to tell you this quick story that happened after I retired. The outstanding Band Booster Club at my oldest daughter's school held an annual Spaghetti Dinner. The event was held to raise money for the band. Since our daughter was a member, I decided to volunteer to help. I was responsible for making the spaghetti sauce from items donated by local restaurants.

It was unbelievable how the local restaurants stepped up, thanks to the Band Boosters. I went through the donated items and selected the ingredients that I thought would make the best-tasting spaghetti sauce. After working all afternoon, I put together a commercial-size pan of spaghetti sauce that I thought would taste good enough for a successful dinner. I knew it was good, but I had no idea how good it really was. I'll explain, keep reading.

I guarded that sauce until dinner started, which turned out to be a blessing. Good thing I did, because a volunteer on the second shift came into the kitchen and went directly to the pan of spaghetti sauce. He looked at the pot of sauce for a few seconds, then he picked up a container of some kind of seasoning and was about to add it to the sauce. I exploded so loudly that he nearly jumped out of his shoes. I really don't know what his intentions were, but I was not having it.

Anyway, the sauce turned out to be a masterpiece. At the end of the dinner, we placed the extra sauce in containers, hoping it would sell. Listen, these boosters didn't miss an opportunity to make more money! Shortly after the program ended, we had the father of one family come in and ask if we had any more sauce for sale. As we handed him a container, he told us that his family insisted he come back and get more.

You should have seen the huge smile on my face. I was very happy because of the way things turned out. The boosters did an outstanding job, the event was a huge success, and yes, we sold all the sauce that day. The committee received an unbelievable number of compliments during and immediately after the event. That made me proud.

Let's fast forward about twenty years or so, and I'll tell you another story about that spaghetti sauce. As I stated earlier, our oldest daughter was a senior in high school when the above-mentioned event happened. Years later, my wife and I attended our grandson's band concert. He is the second child of our oldest daughter and was a freshman in high school at this time.

As my grandson introduced me to his friend's mother, she stated with excitement, "Oh yes. I know who you are! You are the man who made that delicious spaghetti sauce at the Band Boosters' spaghetti supper." My face was all teeth again. All I could do was thank the lady. I remember thinking, "Dang! That sauce must have been better than I thought. Over 20 years later, and this lady still remembered me and how good that spaghetti sauce was."

I absolutely loved it!

CHAPTER 14

My Departure from the General's Staff

Apparently, the General had gotten the word that I had been talking about wanting to leave his staff. Sometimes I run my mouth too much, but this time it worked to my advantage. The General struck up a conversation with me during lunch one day, as he often did. "Henry, what do you want to do when your time is up here?" The question came so abruptly that it kind of shocked me; however, I took total advantage of it.

"Well, sir, I want to become a Drill Sergeant. I think I would enjoy training new soldiers." That was my response without hesitation. I really don't remember what the General said at that point, only what he did. It wasn't long before I learned just how much influence he had on my career.

Shortly after that conversation, I received orders to attend Drill Sergeant Training. This was super great for me because the training was conducted right there at Fort Jackson, where

I was stationed. I thanked the General for all he had done for me, and I thanked his family for being so gracious to me.

As it turned out, this move was my ticket to never leaving the States for the rest of my career. Keep reading. I will explain later.

CHAPTER 15

Drill Sergeant Training and My First Assignment

I entered Drill Sergeant Training directly from the General's Staff. There were a few changes I had to make going from the Food Service field to the training arena. My grade stayed the same (E6), but my rank changed from Specialist to Staff Sergeant. My basic work uniform changed from the black pants and white shirt uniform to the green work uniform (Army fatigues). This was the uniform that most soldiers wore when they worked or were in combat; what you might call battle dress.

One experience I remember from Drill Sergeant Training stands out clearly. The training company was made up of four or five platoons with about twenty soldiers in each. On this day, when the instructor prepared the company to march back to our area, he formed us up in a very tight parking lot. Watching how things unfolded, I believed he placed us there on purpose as a training exercise.

Picture this: the company was formed in an area almost as long as a half city block and about as wide as a two-lane road. The company faced the opposite direction from where we needed to go. On this particular day, one of the student trainees who was in charge of the company was given the responsibility to move us once the instructor gave the command. After trying a couple of times, that trainee leader could not move the company. The instructor then gave another trainee the opportunity, but he also failed.

I don't remember if I was selected or volunteered, but eventually, I got the chance to lead the company. By that time, I had studied what needed to be done to move us out of the lot and back to our company area. I gave the following commands: "Company, Attention! Right Face! Forward, March!" Then I called for two immediate "Column Lefts," which turned the company in the opposite direction. Each platoon leader repeated my commands at the proper time, keeping the formation intact.

Once the lead platoon reached the point where I could guide them toward the adjacent street, I gave new commands. "Column Half Left, March! Column Half Right, March!"

This lined us up with the exit from the lot. As we neared the street, I followed with, "Road Guards Left and Right, Post!" The Road Guards stopped traffic in both directions so the company could safely move into the street. Finally, when the formation was clear, I gave the command, "Road Guards, Post!" to bring them back into the company formation.

At this point, I was in hog heaven. I have never really known what that phrase meant, but man, I felt great. I was the "Drill Cook." I was one of only a few students in the class who was not a combat soldier, and the only one who had worked in another profession within the Army. Since I came from food service, the students gave me that nickname, and it didn't bother me one bit. In fact, I kind of liked it.

As I mentioned earlier, I had no problems with any part of Drill Sergeant Training except the Basic Weapons Training. I found that portion of the course extremely difficult. I had no experience with most of the Army's standard weapons except the M16 rifle. I trained briefly with the M14, but the M16 was only assigned after I finished Basic and Advanced Individual Training. My only exposure to the M16 before Drill Sergeant School came during a quick briefing directed by higher headquarters while I was enroute to Vietnam.

I failed the BRM (Basic Rifle Marksmanship) portion of training. First, let me say, I take full responsibility for failing that test. But I must admit, the weapons training seemed too in-depth for the time allocated. I simply could not remember half of the information on the large number of weapons used by combat soldiers. The training, I felt, was geared toward Combat Arms soldiers, which I understood, but as a Support soldier, it was overwhelming. Several of us, including Army Reservists, had little to no experience with the vast arsenal they expected us to master.

I expected the inevitable outcome after my test, and I was very disappointed. Deep down, I knew I could be a great Drill Sergeant, so I took it upon myself to speak with the school Sergeant Major. I didn't complain about the course. I accepted responsibility and shared my determination to succeed. As it turned out, the Commandant granted a waiver for me and about sixteen Army Reservist trainees to repeat the weapons training only. I was excited to get that second chance, and I smoked that test the second time around.

I graduated with one of the early 1970 classes. To my surprise, the First Lady of Fort Jackson, Mrs. Coleman, attended my graduation and presented me with a gold whistle. That whistle became part of my uniform for as long as I served on Drill Sergeant duty. I cannot express how much it meant that the Commanding General's wife thought enough of me to attend my graduation. It confirmed in my heart that I must have done something right for that family. I'll share more about them later.

CHAPTER 16

My First Drill Sergeant Assignment

I was assigned to Company C, 11th Basic Training Battalion. The company trainees were in their last week of training, and my first duty was to bring up the rear of the company during several miles of marching. They were returning from a week of field training, or bivouac, back to the company area. I was excited but also knew this would be my first real test as a Drill Sergeant.

Quick story: As the troops marched through rough terrain, one of the trainees in the back thought he would punk me, the new Drill Sergeant. The conversation went something like this. The trainee said, "Man, I need a cigarette. Drill Sergeant, do you mind if I smoke?" Man, did he hit a nerve. I was pissed, but I kept my cool and responded, "Are you supposed to be talking, trainee?" "No, Drill Sergeant," he quickly replied.

Everything went smoothly the rest of the way back to the company area. A few hours later, to my surprise, the Field

First Sergeant placed me in charge of the formation while he and the other Drill Sergeants went to the Orderly Room for a meeting. This gave me the perfect chance to address the disorderly trainee. Before I explain what I did, let me clarify the role of the Field First Sergeant. The First Sergeant is the senior Noncommissioned Officer in a training company, while the Field First Sergeant is the senior NCO in charge of the company training activities.

Back to the trainee. The Field First Sergeant, putting me in charge, allowed me to act out some of the things I had heard about Drill Sergeants. Most of the time, a Drill Sergeant goes into an acting mode to enforce discipline. It is a great tool for getting the attention of trainees and making sure they understand the seriousness of the training. This was my first chance to instill discipline of my own and make an example.

I walked out in front of the company and commanded, "Company, attention! Parade rest! Private Trainee, front and center!" Once the private was standing before me at attention, I addressed the entire company.

"Private Trainee tried to punk me, the NEW DRILL SERGEANT, while marching back from bivouac."

Then I asked, "Trainees, are you supposed to talk without permission while marching in formation?"

They shouted in unison, "NO, DRILL SERGEANT!"

I followed up, "Are you supposed to smoke while marching in formation?"

Again, the loud reply came: "NO, DRILL SERGEANT!"

I continued to explain, "Private Trainee asked me if he could smoke while we were marching back from bivouac. I really hate to deny a man the opportunity to participate in his favorite pastime; therefore, I will allow Private Trainee to smoke at this time." While the private was still in the position of attention, I ordered him to light up. Once he lit the cigarette and still held it in his mouth, I ordered him to drop to the push-up position.

Picture this: I bent over from my waist, with my hands coupled together behind my back, and the rim of my Drill Sergeant hat almost touching the back of the trainee's head. I demanded in a loud, authoritarian, and totally unnecessary bass voice that he never disrespect a superior again. I dug into his ass, figuratively speaking, to make the point loud and clear. The whole bit was theater, but it worked to get the company's attention.

My point was to make him aware that what he did was wrong and disrespectful to me as well as to his fellow trainees. First, never disrespect your superiors. Next, never underestimate the new kid on the block. Finally, always follow the rules. Man, did I enjoy that.

My First Teaching Assignment

My first full cycle with new trainees reported about two weeks after my assignment. I was assigned the hand-to-hand combat block of instruction. The first class was a total disaster; I blew it. I knew the subject matter, but presenting it was a different story.

About five minutes into the period, the Field First Sergeant jumped up onto the roughly four-foot-high, raised platform and took over instruction. Watching him teach did everything to help me with that block. He was great, and I followed his lead. I learned that this training was mostly about presentation and poise enough to convince your students that they would definitely kick the enemy's ass in hand-to-hand combat.

My next block and all future blocks of instruction were on target. I realized that knowing the material was only half of the job. What I needed was to see how seasoned Drill Sergeants presented the material and then copy that style. That change made all the difference.

Paying attention to each instructor's delivery of a block of instruction was one of the main things I did whenever I was the assisting instructor. As time passed, I began to add my own twist to my delivery techniques. I gained a wealth of experience during Drill Sergeant duty, as well as in other positions I held during my 24 years of service. As a matter of fact, part of my duties as an Education Specialist during my second career was evaluating instructors during their assigned blocks of instruction. That responsibility tied directly to what I had learned in those early years.

Guard Duty Training was also assigned to me, and I was a hit from the start. I noticed that each instructor would use a short story or a funny piece of information pertaining to the subject matter as an icebreaker. These stories were designed to relax the trainees and put them in a learning frame of mind. The story I always used was "The Guard and the Gorilla." My introduction went something like this:

In a thunderous bass voice, I shouted, **"GOOD MORNING, TRAINEES!"** In unison, they roared back, **"GOOD MORNING, DRILL SERGEANT!"**

I continued, "My name is Drill Sergeant Henry Wilson. Welcome to your first block of Guard Duty Training. Before we get into the meat of today's lesson, let me tell you a quick story about a soldier who once pulled guard duty, just like many of you will one day.

"There was a circus on base that we were ordered to protect overnight. One of the guards was assigned to watch over a dangerous gorilla. The animal was shackled with her arms and legs spread apart, secured to the back wall of the cage, and a muzzle fastened tightly over her mouth.

Because of the way the gorilla was shackled to the wall, her privates were clearly exposed. As the guard walked back and forward in front of the cage in a military manner, he noticed that the gorilla was a female. After a while, the gorilla began to look good to the guard. Because there was no one else in the area, the guard decided to get a closer look at the gorilla. He thought, "What could it hurt? She is securely shackled to the wall." (I'm acting the story out as I go along.)

So, he went into the cage. Once in the cage, he got bold enough to touch the gorilla on her stuff. He reached over, touched the gorilla's stuff, and jumped back. He noticed the gorilla kind of grunted, seemingly in pleasure. So, he touched her again. Again, the gorilla grunted more loudly. The guard began to rub her stuff more vigorously. When the guard was sure that the gorilla was enjoying the attention, he decided to get him some.

Again, the guard thought, "What could it hurt? There is nobody around but me and the gorilla."

He mounted the gorilla from the standing position and began to get it on. After a few strokes, it got so good to the gorilla that she broke loose one of her arms and wrapped it around the guard. After a few more grunts and strokes, she broke loose and wrapped the other arm around the guard. Then both legs broke loose, which left the guard and gorilla all wrapped up and getting it on. As the gorilla grunted louder and louder as they rolled around over the ground, the guard began to call out as loudly as he could for the Sergeant of the Guard to come to his guard post.

As the SOG arrived, he noticed the gorilla had the guard all wrapped up, seemingly to do him bodily harm. The SOG shouted out to the guard, "Move your head over, private, and I will shoot the gorilla."

"NO SERGEANT, DON'T SHOOT HER. TAKE THE MUZZLE OFF SO I CAN KISS THIS BITCH."

Man, did this bring a roar of laughter from the trainees. That set the stage for my introduction to the training. I, as well as the trainees, had lots of fun. After the class introduction, I would proceed to the body of the training.

Each Drill Sergeant was assigned to teach two or more classes per eight-week cycle. All of us took pride in training our trainees. The Drill Sergeant's goal was to have no trainee fail the final test. The final test, called the G3 test, determined if a trainee graduated from Basic Training or not.

Although most Drill Sergeants were assigned one or more blocks of instruction to present, each Drill Sergeant would also assist the lead instructors consistently throughout the eight-week cycle. Each block of instruction would be presented from one to as many as eight hours during the cycle. A single class could be presented in one or more sessions per day.

Earlier, I mentioned that I didn't do well in the Weapons Training during Drill Sergeant School. However, as an Assistant Instructor during Basic Training Cycle, I became extremely effective with each weapon the trainees were taught. The M16 Rifle became the Basic Weapon for soldiers during the Vietnam era. Although I, along with other new soldiers, were briefed on the functions of the M16, I had never fired it before Drill Sergeant Training. I quickly became efficient in disassembling, cleaning, and reassembling the weapon within seconds, and I also became very effective in firing the M16. Let me share a quick story.

After each session of Trainee Marksman Training was completed, all excess ammunition had to be fired before leaving the range. The Drill Sergeants and range staff members fired the remaining ammo. On one occasion, after a firing session, I exhibited my firing abilities, which left no doubt in the minds of my constituents that I was a serious shooter. While firing from the standing position, I knocked down every target at, I believe, a distance of 400 meters on every lane.

Later, I was told that the tower operator automatically laid a lot of them down, but the operator denied it. It was kind of funny, but it wouldn't have hurt my feelings if he did. Just firing the M16 was enjoyable for me. Anyway, I learned a lot about the M16 rifle. It seemed to come naturally for me during my Drill Sergeant duty.

Another opportunity to exercise some of my natural abilities came during the M60 Machine Gun training for the trainees. This was actually training for me as well. Even though I assisted the Lead Instructor, I paid attention to every detail of his instructions. At the end of the training period, I got the opportunity to fire the M60 Machine Gun for the first time. I absolutely loved it. There were different types of targets on this range: an old tank, simulated bunkers, and other disabled equipment that could have been used in war situations.

Just as on the M16 range, the excess ammo had to be fired before the day was over. I grabbed the M60 with a soldier to assist me, and we fed a long string of ammo from the can into the weapon. It seemed like a hundred rounds or more, and I was kind of nervous at first. After I fired about three bursts of three rounds each, I said to myself as I continued to fire, "Man, this is a piece of cake." I could see exactly where my shots were going.

From the standing position, holding the weapon from my right hip like Rambo (Smile), I walked four or five bursts of three rounds each downrange until I reached the tank about 300 meters away. Once I reached the target, I unloaded the rest of the ammo directly onto each of the targets. I have to tell you, it felt really great. I smiled, stepped back, and cleared the weapon. One range staff member who was near me said emphatically, "I KNOW WHAT YOU DID IN VIETNAM."

The staff member assumed that I had served as a gunner in Vietnam, perhaps firing from a combat helicopter. I thought to myself, "Yeah, right." I just smiled at him, walked over, and took charge of my trainees, but I appreciated the compliment. Even with those inherent weapon skills, I am extremely happy that I was not an infantry soldier. Those guys caught hell in

Vietnam, but they also dished out much more hell than they received.

I would never pretend to have gone through even a fraction of what they have gone through. I will always have the utmost respect for our combat soldiers. Sure, I executed my responsibilities when I served in Vietnam, and I was always prepared to engage with the enemy if needed. However, I will never presume that what I went through compares to what our combat soldiers experienced while doing their jobs.

CHAPTER 17

General Coleman's Departure from Fort Jackson

Remember the Base Commander that I worked for? I was surprised but so excited to receive an invitation to attend the Change of Command Ceremony for his departure. It was truly an honor. My Company First Sergeant did not hesitate to give me the time away from my duties to attend. Honestly, with so many city leaders, senior commissioned and noncommissioned officers, and other dignitaries there, I never expected to get anywhere near the General.

The ceremony included a modified vehicle especially designed for the Pass and Review portion of the event. In this case, the departing Base Commander and the Base Command Sergeant Major, the senior noncommissioned officer for the base, stood on the back portion of the vehicle and waved to those along the viewing route. I knew that the driver would carry the General off base through the main gate. Therefore, I decided to go to that location, hoping at least to wave farewell to the General as he exited the base.

As it turned out, I got more than I ever hoped for. As the vehicle approached the gate, the General spotted me as I rendered my hand salute. I noticed him motioning for the driver to stop. To my surprise, he dismounted the vehicle and walked toward me. I immediately moved to meet him, preparing once again to render a salute, but instead he reached out for a handshake.

Let me say this: I always knew this was a real man, a caring man, and an outstanding Commander. But as we shook hands, I also saw the sensitive side of him. He became emotional as he confided in me how much he loved Fort Jackson, how deeply he cared about the soldiers, their families, and the workers on the base. He told me that he really did not want to leave Fort Jackson.

I responded, "But you know you got to go, right, sir?" He smiled faintly and said, "Yes, Henry, I do. Thanks for what you have done for my family and me. I will look you up when I get back from Vietnam."

I answered, "Thank you, sir. I look forward to it."

Although I knew the General was an honest man, I never really expected to see or hear from him or any member of his family again. However, I was proven wrong. Keep reading.

Abruptly and surprisingly, one day several months later, I received a call from the General informing me that he was back in town and wanted to come to my home for a short visit. Of course, I accepted his request with honor. On the evening that he could visit, I was scheduled for Charge of Quarters (CQ) of my Company. Fortunately, I was able to switch that assignment with a fellow Drill Sergeant, SSG Gill, who was very accommodating, so that I could visit with the General.

The General visited me at my home in Family Housing early one evening, where we sat and talked for about an hour or so. He informed me that he returned home early because the position he was reassigned to in Vietnam never matured. Therefore, he was reassigned to Fort Knox, KY, where he would be the Commander of a Regional Ready Army Reserve Component. We talked about my family and me as well.

During the conversation, as I brought him up to date about my job, he mentioned that he had a position on his new staff in Kentucky that he would like to put me in, but it was at a higher pay grade. My response was, "No problem, sir. Just promote me, and I will be happy to accept the position." We both laughed. I was serious, but he didn't bite. I stayed in contact with the General and his family more during my career.

If you know anything about the military, you may wonder why a Major General, a two-star Commissioned Officer, was so fond of a Noncommissioned Officer, a Staff Sergeant (E6). It's like the difference between the mayor of a major city and, perhaps, a mid-level service worker. Well, no problem. For the longest time, I wondered about the same thing. It became even more puzzling after I spoke with his baby daughter years later.

My Drill Sergeant assignment became decreasingly exciting after the Department of the Army went to an "All-Volunteer Army," which changed the way we conducted Basic Training. This was the first of many instances where higher headquarters made changes in training without consulting the people on the ground. One day, as the Drill Sergeants were reviewing drill movements, I noticed a Noncommissioned Officer approaching our formations with a clipboard in his hand. He informed us that we could not practice Drill and Ceremony moves during the present period.

Thinking he was joking or crazy, we dismissed him and kept doing what we had always done. Of course, the Sergeant went directly to the Company Orderly Room. About five minutes later, our Field First Sergeant ordered us to stop training and report to the Orderly Room. During that meeting, we were informed that we could no longer conduct training as usual. We had to follow the script developed by the Department of the Army and approved by Congress.

The first problem with this change was, check this out, there were several blocks of instruction during the newly developed Basic Training that required more Drill Sergeants than were assigned to each company. Depending on the phase of training a company was in at any given time, its Drill Sergeants were loaned to other companies to meet staffing requirements set by the new All-Volunteer Army. This constant movement of Drill Sergeants created confusion and instability within the companies. It disrupted the consistency of training and limited the bonds that Drill Sergeants were able to form with their trainees.

Several times during the first six months of the new training system, I went to work not knowing which Basic Training Company I would be assigned to that day. On many occasions, I found myself working with a company staff I had never met before. It was disorganized and often frustrating, but being the professional that I was, it was no real problem. I did what I had always done. I made it work.

CHAPTER 18

I Got a Break

During the early part of 1973, the Department of the Army ordered the Base Commander of Fort Jackson, South Carolina, to lower its enlisted strength by, I believe, 500 soldiers. The first to go were those soldiers who had been assigned to Fort Jackson the longest. Fortunately, I was in that group. I loved Fort Jackson, but I was ready for a change.

I was reclassified into the Military Police field as a Correctional Officer. My family stayed in Fort Jackson while I attended the Military Police Correction Officer Course at Fort McClellan, Alabama. All soldiers in my category were assigned to a class consisting of about ten mid-level soldiers who had also been reclassified. The first problem we encountered was that we were housed in a dorm on a college campus while attending the course. That was a big mistake, but it became a great learning experience for me.

The students treated us like shit. They called us baby killers, low-life, and any other negative name that entered their minds. Some of the soldiers got upset and wanted to retaliate, but I convinced them that, as representatives of the United States Army, we could not get involved in hostile situations. Because of my leadership ability and resiliency, I was able to help them stay focused. Most importantly, I connected with a few students who were willing to talk with us during meals and breaks. Most of our conversations centered around Vietnam and the military, which gave me an opportunity to share my perspective and, hopefully, change some of their perceptions.

This assignment may have been what convinced me to apply for recruiting duty. I was always talking to people about joining the Army, and I realized that I enjoyed those conversations. People's attitudes are very different today than they were back in the 60s and 70s. Today, I wear my headgear that shows I am a Vietnam Veteran almost everywhere I go. Nearly everyone who notices thanks me for serving. It doesn't matter what race, gender, or age they are. They smile, wave, or shake my hand. My headgear also sparks many different conversations that I truly love.

I have another quick story before I get back to the main timeline. In July 2022, my family celebrated our granddaughter's birthday at a Longhorn Restaurant in Murfreesboro, Tennessee. As we walked toward our table, I passed a fellow veteran, who was white, sitting with his family. As I approached his table, the veteran reached out, gave me a fist bump, and shouted, "Hey, brother. Welcome home." That made me feel so great; it sent chills throughout my body, and I barely held back tears.

As if that wasn't enough, a manager who witnessed the whole exchange between me and the other veteran came to our table, squatted next to me, and thanked me for my service. For the next minute, she talked about the camaraderie she had just witnessed between fellow veterans. The manager left but returned a few minutes later, thanked me again, smiled, and handed me a gift card. I was almost okay until my son asked me what was going on. I was so choked up that it took a few minutes before I could speak.

To this day, I continue to get those "Thank you for your service" comments, but that experience I will never forget. I guess as long as I wear my veteran identifier, the thanks will continue to come. Since that day, I have been greeted with the phrase "Welcome home, brother" by a lot of veterans, especially white vets. I am extremely glad that I lived long enough to witness this change in people's attitudes toward all of our service members and veterans. It is truly one of the most meaningful shifts I have seen in my lifetime.

Back to Military Police Correctional Specialist Training. In spite of some unforeseen disruptions, our training went well. Because we were experienced soldiers, we completed the course about a week faster than the normal cycle. However, because of Army constraints, we could not transfer to our next assignment early. The Commander in charge of the Correctional Officer portion of the training took the opportunity to use us to update the U.S. Army's Confinement Facility Training Manual, which turned out to be an impressive experience for me.

When we were reclassified, each of us was given a training reporting date and a permanent assignment location. I was assigned to Fort Polk, Louisiana, home of Special Forces, Infantry Units, and a Military Police Combat Training

Battalion. Fort Polk was also the home of the U.S. Army Joint Readiness Training. When the guys found out about my assignment, all I heard was how bad the location was. They said it was all jungle, swamps, and endless field training. "Man, you don't want to go to Fort Polk," was all I heard.

Ordinarily, a soldier would not have a choice to change assignments once they were given orders. I remembered that I had someone in my corner whom I never realized I might actually need. Remember the General I worked for? He told me to call him if I ever needed anything. Normally, I never let anyone influence my decisions, but when I had so many soldiers warning me about the same place, I listened and thought about what would benefit my family most.

About a week into training, I called and spoke with General Coleman. We talked about a few things before I got to the point of my call. I knew he would see straight through any bullshit, so I came straight to the point. "Sir, I have orders to Fort Polk, but I would like an assignment to Fort Knox. Would you mind helping me with that, sir?"

The General told me that he would look into it, but he said, "I want you to call me back when you get in the last week of training".

The first day of my last week of training, I followed the General's instructions. About three days before graduation, I was ordered to report to personnel, at which time I was told that my orders had been changed. The conversation went like this.

The clerk: "SSG Wilson, we are sorry, but we received notice today that your assignment has been changed from Fort Polk, LA, to Fort Knox, KY."

"What? Oh man, how did that happen?" Of course, I was putting on an act.

The clerk replied, "I have no idea. I'm just the messenger."

"Oh well, nothing I can do about that now".

"No, but good luck," the clerk said.

I left with a copy of my orders, smiling all the way back to my training area. I told my classmates that my orders had been changed. They wanted to know how I got the orders changed. I just smiled and went about my business. Nobody knew my secret, and that's the way I wanted it, and that's the way it stayed. I will have more about the General later, but for now, back to my training assignment.

As I mentioned earlier, the Commander in charge of Correctional Officer Training used our final week of training to update the Correctional Officer's Training Manual. As it turned out, we ended up changing about 80% of the entire manual. I must say that I was the driving force behind the update. Each day we were covering a pre-selected portion of the manual, which was broken down into four portions representing the four workdays we had left prior to out-processing.

When we completed the selected portion each day, we were off duty for the rest of that day. This is where I upset a lot of my classmates because I was not interested in getting off early. I wanted the manual done right. You see, any job that I am assigned to, I take it seriously. I always believe that you do what you can with what you have when you can. This training manual will be used by thousands of Correctional Officers throughout the U.S. Army. Therefore, it must not be concise. It must be clear and without the need for interpretation. The manual must be written so that there is no doubt about what each officer is supposed to do in certain situations.

The Commander would read each paragraph and explain what it was supposed to mean. Afterward, we would discuss that paragraph specifically, giving our interpretations. By this time in my career, I had become very efficient at interpreting what was written, especially in Army regulations. If I came across a sentence or paragraph that could be interpreted in more than one way, I would not allow the class to move forward. Many times, after class members had already agreed that a paragraph was fine, I would stop them and say, "Wait a minute, can we take another look at this and talk about it?"

I would then proceed to convince them, through a series of questions, that the paragraph could indeed be interpreted in multiple ways. After the second day, the class began to whine and complain because we were spending more time on the project than they wanted. The extra time off was what some of the guys really cared about, but the job at hand was paramount to me. Time and time again, I stopped the class from moving through the manual too quickly. To me, extra time off was not important; what mattered was making sure each paragraph was so clear that no one would ever need to ask for clarification.

It eventually got to the point where the Commander would simply ask, "What do you think, SSG Wilson?" Although I did not like the extra attention, his approach saved a lot of time. In the end, the training manual was completed on schedule and forwarded for approval. During our graduation, the Commander thanked the class members and singled me out with laudatory praise. He called my professional input an outstanding contribution to the project, paramount to his project. As I have stated before, I don't work for awards. However, I can't lie. The Commander's acknowledgment made me feel great.

CHAPTER 19

My First Correctional Specialist Assignment

As with my previous new assignments, although I had no idea what to expect, I was somewhat excited. In August of 1973, I reported to the Military Police Company at Fort Knox, KY, not knowing what position I would be assigned. I need you to remember the next statement. When I entered the orderly room, there was only one soldier who came to greet me. I believe he was the Company Clerk. After I completed the necessary paperwork, he told me to get my family settled and report back the following Monday.

The strangest thing happened the very next day, which was a Thursday. As I processed in through the hospital, there was a soldier lying unconscious on a stretcher in the emergency room. I overheard one of the medics say that they had found the soldier, but he had no ID on him. Being the inquisitive person that I am, I eased over and peeked at the soldier. "I know this guy," I said. "He works at the MP Company," and then I gave them his name.

I had programmed his name and face into my memory bank because he was very accommodating, very informative, and had treated me with respect. The soldier had suffered a serious diabetic attack and could have died without medical treatment. Thanks to that bit of recognition, the medics got the necessary information and began treating him right away. I couldn't help but feel that God put me in the right place at the right time.

Now, back to the statement I asked you to remember. On Monday morning, as I approached the building to report for duty, I noticed a soldier standing in the doorway of the MP Company Headquarters. What caught my attention immediately was that he was a Commissioned Officer. Although I didn't think too much of it at the moment. As I got closer, he opened the door with a big smile and greeted me warmly. "Good morning, SSG Wilson," he said, reaching out to shake my hand.

"I am [his name], your Company Commander. Welcome to the 539th MP Company."

"Thank you, sir," I responded.

His action made me feel great. Later, I discovered the reason for such an unusual welcome. The General I used to work for had called ahead and asked the Company Commander to have me call him when I arrived. That explained why the Commander himself was waiting at the door to greet me. A gesture like that just doesn't happen every day, but I have always believed that all newly assigned soldiers should be greeted with open arms by a member of the company leadership.

Later that same day, I called the General's office as instructed, and one of his staff members arranged an appointment for me to visit. During my visit, we sat and talked for about thirty minutes. We covered a variety of topics: his family, my family, my new assignment, and even a little about life in general. At that time, he was serving as the Commander of one of the U.S. Army Reserve Regional Readiness Groups. The rapport between us was amazing. We were two soldiers, as far apart in rank as the distance is from California to the capital city of Washington, DC, yet we were connected by mutual respect.

I reported to the Confinement Facility and immediately replaced an NCO who was one pay grade higher than me at the time. I became the Guard Commander of one of three prisoner guard platoons at the Fort Knox Confinement Facility. I began my new assignment right away. I had never worked in a confinement facility before, but based on my performance, no one could tell. I must admit, having helped rewrite the Confinement Training Manual really prepared me. Of course, my leadership, management skills, people skills, and desire to get the job done didn't hurt either. (Smile.)

I managed a platoon of energetic Army Correctional Specialists. We were responsible for the custody and control of pretrial prisoners as well as those who had been adjudged but sentenced to sixty days or less. Our duties also included escorting prisoners to and from medical and legal appointments, including court proceedings. My soldiers followed my instructions and guidance extremely well. As all great leaders should, I took care of those soldiers who performed well, followed my lead, and showed loyalty. I supported them in their professional responsibilities and their personal lives whenever possible.

We worked rotating swing shifts, with one day off after each of the first two shifts. After completion of the three rotating shifts, we had three consecutive days off duty. After I was in the Guard Commander's position for a couple of months, I began rewarding high-performing soldiers with an extra day off. What I didn't know then was that I should have placed those soldiers on official pass and reported my action to my immediate supervisor. That step would have protected them in the event of an unfortunate mishap. Fortunately, there were never any issues with the three-day passes while I was in charge.

Unfortunately, one of my soldiers was killed in a car accident while returning from vacation. Because the soldier had been on official leave, there were no complications. Still, one of my Staff Sergeant counterparts had the nerve to say that if I weren't such a "slave driver," that soldier would still be alive. I ignored that comment because all I ever did, and would continue to do in every leadership role, was emphasize the importance of being where you are supposed to be, on time. I have never apologized for that expectation and never will. It is one of the core principles that kept our unit disciplined and effective.

I taught my soldiers that respect matters, even in a prison environment. I insisted that although inmates were confined within our prison system, they were never to be referred to as "cons." Every prisoner was to be addressed by their rank and last name. I did not accept excuses or shortcuts. My philosophy was simple: do your job, maintain professionalism in both appearance and behavior, and you will be rewarded. I always gave special treatment to those Guards who earned it.

Over time, several incidents required my direct attention. I want to highlight three of those particular experiences.

One of the things I always did during each shift was routinely check each area where prisoners were assigned to work within the confinement facility, several times each day. One afternoon, as I walked through the kitchen, I noticed the cook preparing mashed potatoes using a large commercial mixer. A few prisoners stood nearby watching him work. Just as I approached the cook, an oil seal broke in the mixer and poured about a cup of black oil into the bowl of potatoes. What shocked me most was that the cook began to manually stir the oil right into the potatoes.

Reacting instantly, I absolutely exploded. "You must be out of your fucking mind if you think you are going to serve those fucking potatoes to my soldiers! You need to throw out that shit and come up with something else to serve."

The prisoners froze where they stood, and the cook looked like he'd seen a ghost. I continued, "If you don't have any instant potatoes, you'd better speak with your Mess Sergeant. I'm sure he'll help you out." Using my food service experience, I gave the cook guidance so that lunch would still be served on time. I didn't mean to scare the man so badly, but his actions infuriated me.

I have always been extremely passionate about food preparation, and God surely led me into that kitchen at the right time. I can't imagine how many staff and prisoners would have been impacted if those tainted potatoes had been served. My intervention probably prevented food poisoning on a massive scale. Before leaving, I suggested that the cook inform the Mess Sergeant of the mishap immediately. That incident reinforced to me that leadership sometimes means stepping in hard when something critical is at stake.

The second incident took place during our weekly supply and laundry exchange for the prisoners. During this process, Minimum, Medium, and Maximum Custody prisoners were brought out in groups to exchange their laundry and draw supplies. I happened to be patrolling the yard while the Minimum Custody prisoners were assembled. To my shock, I noticed one of the prisoners smoking. Every prisoner knew that smoking was not allowed while in any formation.

I calmly approached the soldier, hands interlocked behind my back. "Private, who gave you permission to smoke in my courtyard?" I asked.

Boldly, the prisoner replied, "I DID."

His defiance caused the other prisoners who overheard him to laugh out loud. To maintain order, I commanded, "At Ease," quieting the group. I then directed the prisoner to put out his cigarette, which he did, before tossing the butt on the ground. His attitude told me he was trying to piss me off. However, I remained calm and ordered him to pick up the butt and follow me. With one of my guards walking close behind, I escorted him to the maximum custody cells.

Once there, I reassigned him to Maximum Custody, where he remained until his court date. My after-action report documented the incident in full detail and was approved by the Commander. Although writing reports was a requirement by Army Confinement Regulations, in this case, it turned out to be a critical safeguard for me.

During the prisoner's trial, he was convicted of the charges he was already facing. However, as soon as the judge pronounced him guilty, the prisoner began crying profusely and shouting that I had beaten him inside the Confinement Facility. The judge immediately called a recess and ordered a full investigation into his claim. The Senior NCO assigned to investigate questioned me thoroughly, asking repeatedly if I had touched or struck the prisoner in any way. My answers were consistent: "No."

As the Sergeant wrapped up his investigation, he asked rhetorically, "Why would the prisoner make those accusations if they weren't true?" I replied sharply, "If you read my After-Action Report thoroughly, you'll see exactly why. The prisoner had a plan to provoke me or one of my guards into violating his rights, hoping the judge would show him leniency during sentencing. That's why he was so bold in his behavior that day." Because I followed regulations to the letter, the investigation closed in my favor.

The third incident happened about mid-evening one night. I, along with team members, confined a prisoner who had been AWOL for quite some time. During in-processing, there are several things we must do. This includes searching the prisoner and their belongings. This prisoner was very cooperative until the guard asked for the stack of letters he held very securely in his hand. The prisoner became extremely agitated and hostile. At this point, I stepped in and took over the process. I talked to the prisoner in hopes of convincing him to hand over the letters, but he did not budge.

He was dead set on not giving up the letters because, as he said, they were extremely personal to him. I informed the prisoner that I would only search each envelope and return all of them to him. He continued to refuse to hand over the letters. I had one more option, which was part of the control process that required the Confinement Facility Commander's approval. In the control center, there was an emergency closet that contained riot equipment, gas masks, police batons, and additional first-aid supplies that should be used during emergency situations only. The Guard Commander (me) must contact the Confinement Facility Commander, who would make the decision to open the closet or not.

Once I explained the situation to the Confinement Facility Commander, I went on to say, "Sir, I am requesting authorization to use the necessary emergency equipment to combat this problem. However, I will exercise 'show of force' first in hopes that he will give up the letters."

"What do you think I should do, SSG Wilson?"

I was really surprised by his question, but I responded emphatically, "I think you should authorize the use of the Emergency Equipment, sir."

The Confinement Facility Commander stated, "Do it," and hung up the phone.

The action by the Commander showed the trust he had in me. I felt good about that. While three of my guards kept the prisoner secure, I retrieved four batons from the closet and handed one to each of the guards, instructing them to follow my lead only. As a psychological move, I made sure the prisoner could see what I was doing. After handing a baton to each guard,

I walked to within a foot in front of the prisoner, looked him in his eyes, and stated, "Private. Will you hand over the letters?"

"No."

I moved closer to the prisoner, changed my expression to seriously ugly, and stated, "YOU WILL GIVE UP THE LETTERS, private. ONE WAY OR ANOTHER, YOU WILL GIVE THEM UP."

It was a psychological line of information that I had used several times during my career. It worked every time, including this time. The prisoner paused for a moment. Then he seriously looked at me and each of the guards, who were showing anxiousness with an expression almost as ugly as mine. He turned back to me, showing fear in his eyes, and stated, "You are going to give them back to me, right?"

"I promise," I said.

The prisoner complied, and I had the guards complete the process and put him to bed for the night. Shortly after the prisoner was locked in his cell, I checked on him and told him to have a good night.

"Thanks, Sergeant," he replied and turned over for the night.

Once the Commander read my report the next day, he thanked me and congratulated me on a job well done. He then asked exactly how I convinced the prisoner to turn over the letters. I told the Commander that I was confident I could convince the prisoner to give up the letters with a simple show of force. You see, I had used the same technique before on an intoxicated Marine at the NCO Club, where I worked as a Master at Arms. Let me explain.

During my off-duty time working at the Fort Knox Noncommissioned Officers' Club Security, we encountered an unruly Marine. Apparently, the Marine had become hostile after having too much to drink. As I worked in one of the small pubs of the club, an announcement came through the intercom calling for a designated MA to report to the main ballroom. I arrived in the ballroom area, where a few Army soldiers and a Marine were having drinks. The Marine obviously seemed intoxicated.

When I arrived, the lead Master at Arms was trying to convince the Marine to leave the club. The Marine refused, saying, "I ain't going no goddamn where." The crowd around the Marine seemed to give him courage as they laughed. I stood about four feet from the Marine with my arms folded while I surveyed the crowd. The lead Master at Arms told the Marine to leave the club, or he would have to call the MPs. The Marine stated, "Call them. I don't give a fuck."

After the lead MA left the area to call the MPs, I slowly stepped forward enough to lean over toward the Marine with my arms still folded and whispered in the Marine's ear, "Man, why don't you save yourself some trouble and just leave?"

The Marine, in a loud voice, stated, "I ain't going no god-damn where." Once again, the crowd cheered him on. I straightened up, arms still folded, and made sure the Marine saw me change my expression to seriously ugly.

I moved closer to the Marine, looked him directly in his eyes, and stated, "Oh, you are going to leave." With emphasis, I added, "One way or another, YOU ARE GOING TO LEAVE THIS CLUB!!!" The Marine looked around at the crowd, looked back at me, took a last drink from his glass, slammed the glass on the table, got up, and left the club. As I followed him through the crowd to the exit, the crowd's attitude seemed to change from cheers to, "Look at that MA. He thinks he's bad." I got to tell you that I did have a proud strut about my walk that indicated I would not take any shit. Anyway, back to the serious stuff.

I always knew that the Confinement Facility Commander and the Noncommissioned Officer in Charge were very pleased with the way I did things. What they did next really shocked me. There was a new Staff Sergeant Correctional Specialist assigned. Since I was junior in grade to the SSG, he was supposed to replace me as Guard Commander. The Facility Commander was not having any of that. He talked with the SSG and explained that I was the Guard Commander and that he would be working under my supervision. That was that. Surprisingly, the SSG had no problem with working as the second man on my guard platoon. I continued to do what I did. At the same time, I gave him the respect he deserved. We worked well together.

A few months later, I became Chief of the "In and Out-Processing Section." My staff consisted of three additional soldiers: one Sergeant (E5) and two Specialists, 4th Class. My team was responsible for processing all prisoners in and out of the Confinement Facility during normal business hours. We also processed prisoners who were confined during Friday evenings and weekends. However, we processed those prisoners on the facility's first business workday after confinement.

During the first week in this position, I studied the Army Regulations and Confinement Training Manual intensely. Within two weeks, I trained my team on those regulations as well. I trained them on the use of every document and procedure we had to use to fulfill our responsibilities. The most important document we used was the "Prisoner History Form." Because of the importance of this document to the Confinement Facility and the prisoner counselors, I demanded that my team members complete this document free of errors. I did not accept any completed form with white-outs or erasures. I felt that there should not be any questions about a prisoner's status when he or she was confined.

To ensure this happened, I checked each form before sending it forward. I also realized that the information revealed a lot about the prisoner's personal and military life. Based on that alone, I thought the form should be completed by the Prisoner Counseling Section Team. It appeared to me that the questions on the form were intended to aid the counselors in gathering initial information that would help them support the prisoners during confinement. As my section continued to complete the form, I also continued to analyze the information we got from each prisoner. Once I was convinced of my findings, I briefed my team members on my intention. I realized that this section could be set up to run more efficiently, which would benefit the Confinement Facility, the staff, and the prisoners.

Once I completed my research, I requested a meeting with my boss, who was the Noncommissioned Officer in Charge of the Confinement Facility, and the Chief of the Prisoner Counseling Section. During the meeting, I got straight to the point. I suggested to our boss that the Prisoner Counseling Section be responsible for completing this form. During my briefing, I pointed out and showed justification that the form was meant to be completed by the person responsible for gathering sensitive personal and military information from the prisoner.

Since the information was extremely important to the treatment of the prisoners, and since each counselor is assigned a certain number of prisoners, they should start completing the form during their initial contact with the prisoner. In addition, the form should be updated each time the counselor makes contact with the prisoner. The form will ultimately be maintained in the prisoner's confinement file. The information contained within the file will help the counselor stay updated with the prisoner's confinement history, thus giving them the necessary information that will help the prisoner get through their confinement period successfully. In addition, it will allow anyone who has legal access to the prisoner to have all they need to know about the prisoner.

As I passed my findings on, it was obvious that the Chief of the Prisoner Counseling Section was becoming agitated. He became so upset that he actually lost it for a moment. "At ease, SSG," he shouted to me. "Who are you to tell me how to run my section?" he continued.

I said to the Section Chief in a factual but respectful tone, "First of all, SFC, we are both Section Chiefs in this Confinement Facility. As long as I am in this position, don't ever try to pull rank on me again. I am just trying to put your section, as well as my section, in line with what's required in the Training Manual. Based on what I know and what I have seen during the last 18 months, the training manual has a great plan. In fact, it's the law. We just need to get on board with Army Confinement requirements."

The more I talked, I could tell the Chief was getting increasingly pissed, but did I care? No, because I knew I was right. At this point, our boss chimed in, "SSG Wilson is right, SFC. We need to take a good look at this." I also requested through my boss that I be allowed to move my section to the building inside the compound. I reminded my boss that the facility was already built according to regulations, but was never used accordingly. I went on to explain that the outside fence, inner fence, and the building that I recommended moving my section into were already in place.

The building is inside the compound with no chance for prisoners to escape. It also has bath facilities in place. I have done an extensive check of the facility and the building that I want to move into. As I mentioned, the facility is already built according to Army Regulations. All I would have to do is make a few minor modifications to the receiving area that would be in keeping with my section's responsibilities. My boss gave approval for both plans, with the Facility Commander's blessing, of course.

I briefed my team on the changes and immediately started plans to relocate my office. I met with key people whom I needed to help me get set up in my new location. My purpose was to introduce them to my plans so that they could confirm which jobs each supervisor could help me with. This included my design for the processing stations to process up to 16 prisoners at the same time.

This also included painting, plumbing, etc. Once everybody was on board with my plan, I moved forward like a charging bull. One of the lieutenants in my management chain questioned whether I could get the changes completed by the next dignitary briefing, which was in just a few days. Once I gave the lieutenant a synopsis of my plan, he just didn't see how I could get the changes completed prior to the briefing. I convinced him that I had everything under control. You see, we had a high-ranking official visit the facility every Tuesday.

He or she was escorted throughout the facility by the Confinement Facility Commander or his representative. I asked the lieutenant not to worry because I would be ready to go on Tuesday. In just three days, I set my office up, adjusted the processing Standard Operating Procedures, and prepared the Prisoner and Visitors Briefings. I was ready to go the following Tuesday.

I gave my first briefing in the new space when the visitors came through. My Confinement Facility Commander was full of compliments and smiles. My actions caused several sections within the confinement facility, and especially my section, to be in concurrence according to the training manual. Let me explain.

When prisoners were transported to the Confinement Facility, the transporting vehicle was driven through a gate into a fenced-in area called the holding area. That area was large enough to accommodate a twenty-four-passenger bus. Once the vehicle was inside this area, the Correctional Specialist locked the outside gate. The inside gate stayed secure until all MPs, including the driver, had their credentials checked and the prisoner confinement orders were verified.

At this point, the prisoners were ordered out of the vehicle and searched by the Correctional Specialists. Once they were cleared, the inside gate was opened, and my section took charge of the prisoner. Once the inside gate was secured, the outside gate was unlocked, and only then were the MPs allowed to leave.

The prisoners were taken into our new processing area. Each prisoner was ordered to strip and then thoroughly searched. Their belongings were thoroughly inventoried and secured. Each prisoner was required to shower and was given a thorough briefing of their dos and don'ts while in confinement. Each prisoner was finally assigned to a cell.

The "In and Out-Processing Section" worked extremely well. I felt good about what I had done for the facility, including forcing the Prisoner Counseling Section to take over their responsibilities.

Just before changing jobs, my wife and I brought our second lovely child, a girl, into the world. It was an all-day delivery. I had worked the midnight shift and was scheduled to get off duty at 7:00 a.m. I guess our baby Dana had other ideas. (Smile.) My wife called around 5:00 a.m., telling me that she was ready. I went home immediately and took her to the hospital, only to be told that my wife needed to go walking to help with the delivery.

Finally, we were placed in a room while she went through labor. Man, was I tired! I lay my head on my wife's stomach as she comforted me with her hand on my face. The only problem was that each time she had labor pains, she would grab my ear and pull and squeeze. The only thing I could do was go where she pulled. Man, did it hurt, but I went with it. (Smile.) Finally, in the early afternoon, our lovely daughter Dana (my mother named her) was born.

Dana was born on June 9, 1974. To this point in her life, Dana has earned her Bachelor of Arts Degree with a concentration in Criminal Justice from the University of Tennessee in Knoxville, TN. She has also earned her Master of Public Administration Degree from Tennessee State University. Dana is doing well in her job, where she is the Senior Business Process Manager for Holland & Knight Law Firm. She is married to Luis, a wonderful young man from Mexico City who is now a citizen of the United States of America. Luis and Dana gave my wife and me a beautiful granddaughter, Natalia Carlee Morales (her middle name, Carlee, was taken from her two grandmothers).

I have to tell you this quick story. My wife and I had planned to have only two children. We were hoping for a boy and a girl, but God blessed us with two lovely girls. Well, because we decided that I should get a vasectomy, we were required by Army Regulations to go through counseling. There was a waiting period of six months after counseling, which turned out to be a blessing for us. Six months later, on the day of the appointment, my fear of the surgery got the best of me. (Smile.) I called the hospital, and my exact words were, "You have an appointment for SSG Henry Wilson scheduled for 1:00 p.m. today."

The attendant said, "Yes, we do."

I replied, "He ain't going to show. Thanks." And I hung up the phone. I'll talk more about our blessings later.

I was extremely successful in my assignment as Chief of the In and Out-Processing Section of the confinement facility. As in previous assignments, I continued to work to accomplish the mission of the section. My superiors never worried about any areas of the facility in which I was involved. I received many laudatory congratulations as well as written commendations.

During my assignment to the Confinement Facility in Fort Knox, the Facility NCOIC approached me with a recommendation that I could not turn down. The Master Sergeant thought it would be a great idea for me to apply to become a Master Mason. He explained as much as he could about the fraternity, which was enough to lead me in the right direction. I applied right away and was accepted for initiation. After a few months as a Master Mason, I was voted in as Junior Warden and later as Senior Warden of the local chapter. I really enjoyed doing great things for the people who lived in the communities in and around Fort Knox.

The General's Baby Daughter's High School Graduation.

Quick story: Remember the General I cooked for at Fort Jackson, SC? Well, sometime during my assignment at Fort Knox, I received a high school graduation invitation from his youngest daughter, the 13-year-old I told you about earlier. As my wife and I sat in the audience, I took a few videos of the ceremonies, specifically of the graduating seniors. Since we were guests of the General, we had great seats right down front. As Coleman's daughter approached the stage to receive her diploma, I stood in a position in front of the crowd to get a picture as she received it.

As she departed the stage, she noticed me taking her picture. She stopped, screamed, ran to me, and gave me the biggest and longest hug ever. As I sat through the rest of the ceremony, again I wondered what I did to cause this family to become so crazy about me. Just being me, I guess, and treating people with respect and honesty. That's why you should always do the right thing for people, because you never know how you affect their lives.

I spoke with the General's wife not long after I transferred from Fort Knox, at which time she told me that the General had lost his battle with cancer. The news was hurtful, but I understood that each of us is going to depart this earth one of these days. That's the way God set his world up. I'll have more about the family later.

CHAPTER 20

My U.S. Army Recruiting Experience

In early 1976, there was a push to select as many Army Recruiters as possible to fill the soldier shortage throughout the U.S. Army. Once I read the announcement, I spoke with my wife about applying. She agreed that I should because I was always talking to people about joining the Army. So, I applied for the job right away.

The Recruiting Command screened, if I remember correctly, the records of over three hundred applicants from Fort Knox alone. Roughly two hundred of those soldiers were selected for an interview. Of those interviewed, I was in the group that attended the recruiting course. To be honest with you, I didn't know anything about what it took to be an effective Army Recruiter. I did know, however, that I am an "I can do this" kind of person. So, I never felt fear. It turned out that recruiting was another one of my successful career fields. There was a period when I struggled early on during this assignment. Keep reading. I'll tell you about it later.

After completing the Army Recruiting Course in mid-1976, I was assigned to the Jackson, Mississippi Recruiting Battalion with my duty assignment in Jonesboro, AR. Now, y'all, although Jonesboro was only eighty miles from Memphis, I had never heard of it. Hell, being a country boy from the small town of Millington, TN, I barely heard of Memphis. (Smile.)

Anyway, my wife and I took a pre-assignment trip to check out the town. As we drove into Jonesboro through the suburb of Nettleton, my wife stated with doubt in her voice, "Henry, they don't have sidewalks here." As we drove a few more blocks, she said, "There is a field of goats out there. Henry, I can't live here." Then my poor baby literally started crying. Personally, I had no feelings about the town because the assignment was a done deal for the family and me.

As it turned out, the town was a great assignment for my family and me. As a matter of fact, I am sitting in the living room of one of the first people we met as I write this portion of my book. Deborah Agnew remains one of our best friends to this day. I'll talk more about Deborah later.

During the pre-assignment visit, my wife and I checked on some apartments and schools. I also had the opportunity to visit the recruiting station, which I would share with two other recruiters just to introduce myself. I got to tell you, I was not impressed with these guys who were white. I'll explain why I pointed out their race later. Anyway, during the next week or so, I officially signed into the company. I met my Company Commander and First Sergeant.

The First Sergeant was a piece of work. He was aggressive as hell, but he knew recruiting. Most of all, he knew his company area and his recruiters. The company area extended from Dunklin County in the Bootheel of Missouri to the north and extended through the counties that border the Mississippi River in Eastern AR. It also covered the towns and cities past Helena, AR, and Clarksdale, MS, to the south in southwest Mississippi.

The Company Headquarters, which was the office of the Company Commander and First Sergeant, was in Craighead County. My Recruiting Station covered Greene and Clay Counties to the north, both of which had an all-white population, so I was told, which meant nothing to me as it relates to recruiting. Our station also covered Poinsett and Cross Counties in Arkansas to the south. We were also responsible for Craighead County, which was the home of our Recruiting Station, which was located in Jonesboro. I provide this information because it will help you understand my stories that were created because of this assignment.

On my first day of work, I met with the First Sergeant, who briefed me on my responsibilities as a recruiter, and he also talked to me about the recruiting station and our area of responsibility. In addition, he assigned two southern counties, Poinsett and Cross, to me. Once I completed research on the two counties, I understood why they were assigned to me. These counties were populated with a large percentage of Blacks. The Station Commander, my immediate supervisor, didn't have much to say to me, which I found out later was a good thing. I'll talk more about him a little later.

Nobody mentioned that there was a Black female Army Reserve Recruiter assigned to the station. It wasn't a big deal to me, but I just wondered why she was not mentioned during my introduction. I was told that I was her replacement. Again, I wondered why I was replacing a reserve recruiter. As time passed, I assumed it was a race thing. The First Sergeant had me ride with the reserve recruiter as she went to appointments, supposedly to introduce me to the Jonesboro area. The reserve recruiter disappeared as quickly as she appeared. I only saw her that one time. No big deal. I went forward with my work.

As I went forward, setting up my office area and learning my recruiting area, I spent time with my recruiter counterpart. He and I talked a lot during the next few weeks. One thing he shared with me worth mentioning is that the Station Commander was not an honest person. He said that the Station Commander sat in the office most of the time. The recruiter told me that the Station Commander would often steal his applicants. So that you understand the process, the names of applicants for enlistment had to be called in to the Military Entrance Processing Station (MEPS) by the Station Commander. The recruiter told me that when he would call his applicants' names in for processing, the Station Commander would add his own name as the recruiter of record for credit. This is extremely important because each recruiter has an assigned number of applicants to enlist each month.

I asked the recruiter why he would allow the Station Commander to do that. The recruiter told me that he spoke with the Station Commander about this problem. He said that the Station Commander told him not to worry about it, that it would all work out. I thought to myself, "YOU DUMMY."

I really don't remember having much interaction with the Station Commander. I mean, he was the Station Boss. I expected him to be more involved with me. As I mentioned earlier, the First Sergeant briefed me on the company layout, my area, and our mission. I found out later why. This station was not successful and had not been successful in quite some time. My initial impression of these guys was on point. Keep reading, and you will understand why.

Anyway, I moved forward, trying to put people in the Army. My very first applicant was a prior-service Marine who had been medically discharged from the US Marine Corps. After prequalifying the applicant, I studied the information using the Army Recruiting Regulations. I informed the applicant that he would need approval from our Commanding General to enlist in the U.S. Army and that the process may take a few months. Of course, I explained that it was because of his medical discharge from the Marine Corps. I also explained that if he was serious about enlisting in the U.S. Army, I would do what was within my power to help him enlist. I also explained that I would need his dedication to complete that process. He agreed, and I moved forward with the process.

As I moved forward, putting the packet together, the Station Commander shocked me. He actually spoke to me. "Sarge," he said, "you are wasting your time messing with that guy. They will never approve that waiver." I just stared at him, not saying a word. I thought it was best not to say what I wanted to. You see, unless the regulations say, "not qualified" and "no options," the recruiter does not have the authority to deny processing. I continued doing what I was doing. It took about two weeks for me to put the packet together while I continued my recruiting responsibilities. I forwarded the waiver packet to the appropriate commanders for their approval. I'll have more on that later. Keep reading.

During my first month on mission, around August 1976, I enlisted four or five applicants into the Army. Man, did I feel great. Then everything fell apart. During the next three months, I could not turn over a rock with a bulldozer. I couldn't close a sale to save my life. Man, was I depressed. The First Sergeant was deep in my ass. At one point, he stated, "You've gone up there and gotten just like those other dip-shits in that station." He cursed a lot. He continued, "You had better get your shit together, fast."

Just after he met with me, there was a three-day Company Meeting scheduled in West Memphis, Arkansas. Before attending the meeting, I told my wife that I could not take it anymore, that I was going to quit, and if she wanted to, she could stay on until I got set up in a new assignment. Well, as it turned out, that never happened. During the opening statement at the meeting, we were informed that the Department of the Army had frozen all recruiters who were presently on recruiting duty. There was no way out. All I remembered about the meeting from that point to the end of the meeting was the goat that the First Sergeant was trying to BBQ. I don't remember anything else that was said or done. I remember sitting there trying to figure out what I was doing wrong and what I needed to do to make it right.

When we returned to the office, I informed my counterpart, "Hey man, you probably won't see me much after today. I am taking one of the cars (we only had two cars for three recruiters), and I probably will not come into the office except to replenish my supplies."

I told him to call me at home in the evenings if he needed me. I packed my car with everything I needed to put applicants in the U.S. Army. During the next few months, I established a recruiting program that the area had not seen in that area ever. I continued to use the school method of recruiting, but I needed to put my own spin on my recruiting methods.

I set up Points of Contact (POC) in all the key sites within my area. I visited each of my schools, no matter what the school population was or what area they were located in. I found that the problem with a lot of recruiters was that they stereotyped people based on status or where they lived. Not me. I recruited from every inch of my area. I treated each of my applicants and their parents, guardians, or spouses with respect, and I used them all in my recruiting efforts. A lot of them actually recruited for me.

One quality of salesmanship I learned was closing the sale. Every opening I saw during a sales presentation to ask for a commitment, I went for it. I also learned to watch the body language of the applicant and/or the person who was going to make the decision. That would tell me if I could possibly get a commitment or not.

I have to tell you this quick story. Once, when I was sitting in my car in front of a retired Army Sergeant Major's business, a young high school student approached my car as I sat in the car, going over my notes. She spoke and told me right off that she wanted to enlist in the Army. She was 16 years old and was between her 11th and 12th grades of high school. I informed her that I would contact her once she turned 17 and became a high school senior.

The young lady informed me that her grandmother was her legal guardian. When the time was right, I called her as promised. She informed me that her grandmother was not going to let her join the Army. I spoke with the grandmother by phone and asked for an appointment. She said, with an attitude, "For what? I am not letting my granddaughter go into no Army! You will just be wasting your time."

I responded, "That's fine, ma'am. I just want to come by and talk with you and give you a little insight about our United States Army."

She responded, "You can come over, but my granddaughter ain't going in no Army."

When I visited the grandmother, she was sitting on a screened-in porch of a small, well-kept house with one of her friends. Both were well-dressed, attractive ladies who seemed to be in their mid-sixties to mid-seventies. I could see the granddaughter behind the screen door, but she never came out. I purposely never acknowledged she was there. Instead, I focused my attention and, I must say, my charming smile, on the two ladies.

I introduced myself, making sure to use my military title in the introduction. Both ladies presented their objections to the granddaughter joining the Army. Of course, most of their objections were old myths passed on over the years. Other objections were simply the lack of good information about the U.S. Army. With the experience I gained while working in several career fields since being drafted, together with my ability to compare Army jobs with civilian jobs, I had a lot of insight.

I never sugar-coated any information I presented to the ladies. About halfway through my visit, I noticed that both ladies began to smile increasingly, and they opened up to me more as well. When they began to ask more questions, I knew that was a good thing. In addition, they passed on several names of potential prospects. My intention, if possible, was to let the granddaughter have credit for those prospects after she enlisted.

Yes, there was a point in our conversation that I felt confident I was getting through to not just the grandmother, but her friend as well. As I prepared to leave, the grandmother stated, "She can go if she wants to." I stated, "No ma'am, not now. I would like you to please think about the information I have presented to you. If you have any more questions, please call me. Otherwise, I will contact you in a few days. If you are still in agreement, I will come by and complete the paperwork to enlist your granddaughter into the Delayed Entry Program (DEP)," and I emphasized that graduating from high school was of paramount importance.

A few weeks after the young lady started her senior year of high school, I processed her into the DEP. She departed for training shortly after graduation. I have no doubt that the information I presented to the two ladies was passed on to other members of the community, thus supporting my recruiting efforts. After this experience, I realized that my focus during the meeting with the grandmother and her friend was to take this opportunity to aggressively defend our U.S. Army by dispelling stereotypes that a lot of people have about the military. Since these two senior citizens, I felt, had a voice in the community, I believe I was successful.

During the next several weeks, I built a DEP that was so smooth that our Company Commander never noticed. Perhaps he was listening to the Station Commander's unfounded complaints as he tried to cover up his own deficiencies. At one point, the Company Commander told the First Sergeant that the Station Commander said I never reported to him, that he never knew where I was or what I was doing. The First Sergeant stated, "What the fuck does he mean he doesn't know where SSG Wilson is or what he is doing? All he's got to do is look at the fucking Daily Progress Report (DPR). Hell, SSG Wilson is the only one putting anybody in the Army in that damn station. Leave that man alone."

I got to tell you, it seemed like every other word from the First Sergeant's mouth was a curse word. Not long after that, the Company Commander visited our office, complaining about the lack of production within the Delayed Entry Program (DEP). "Y'all's DEP sucks. Y'all are not working the program."

I paused for a moment, then stated, "Sir, I have people in the DEP."

At that point, I had several applicants enlisted in the DEP, which included the first applicant I processed for enlistment. I am talking about the Prior Service Marine that my Station Commander told me the U.S. Army Recruiting Command Commander would never approve the waiver for. It seemed that the Prior Service Marine couldn't thank me enough for getting him back into the military.

That is why you should never take what someone tells you regarding what someone else will or will not do as factual. If you are not the decision maker, shut up and send the information to the person who has the authority to make the decision.

When I pointed out my success with the DEP to the Company Commander, all he could do was grunt. I thought, *man, this guy must not be reading the right reports.* He gets a production statistic report on each of his recruiters every day. My thought was, what does he do with the report? He surely doesn't read it.

The Station Commander and the other recruiter were sent back to the regular army. I told the Company Commander that I wanted to be the Station Commander. But no, they brought in another guy. Through their actions, I realized that they wanted a white Station Commander. To be more specific, they wanted a Salt and Pepper Team in the station. Their stereotypical minds thought that recruiters would be more successful recruiting their own race of people. Boy, did I prove them wrong!! Keep reading.

The Recruiting Command had a Sergeant First Class (SFC) scheduled to take over the Jonesboro Station. I found out that my Company Commander called that recruiter three times to try to determine if he was black or white based on his accent only. I told you the Company Commander was kind of different, right? Check this out. After he could not tell what race the recruiter was during their first two phone conversations, the Company Commander decided to ask the recruiter straight out if he was black or white.

The recruiter told him, "I am black. Is that a problem?"

The Company Commander told him, "Oh no." However, it seemed kind of strange that the recruiter was abruptly assigned to another recruiting station out of our area.

That recruiter was happy to call me to let me know what happened. Yeah, you guessed it. In a week or so, a white recruiter was sent to Jonesboro to take over as Station Commander. Don't get me wrong. I have no problem with a racially diverse team in every station. What I do have a problem with is that I was well qualified to be the Station Commander. However, it appeared that they wanted a white Station Commander.

I knew that the city of Jonesboro was heavily populated by the white race. As a matter of fact, our county, directly to the north of Jonesboro, had an all-white population, so I was told. None of that mattered to me. I was extremely confident that I could make the station successful. No problem, I kept doing my job and let the Lord make it happen. Keep reading.

Although I had asked to take the position, I was not bitter about them sending in someone else to manage the station. I often use adversity to my advantage. I kept doing my job. I continued to be extremely successful in enlisting people from all over my area. One day, a few months after the new Station Commander was assigned, I overheard the First Sergeant state with excitement, "Wilson is putting white and black people in the army."

That statement confirmed the perception I had formed based on several of my leaders' actions and statements. Although I have always been a great listener and I see more things than most people, it didn't take a rocket scientist to figure out that they wanted a Salt and Pepper Team in the Jonesboro Recruiting Station.

Sometime during the next few months, I was promoted to Senior NCO status (SFC, E7). I had been selected a year earlier, but had to wait for my selection number to come up before being promoted. Let me explain. Each year, the Army selects thousands of soldiers for promotion to the next higher grade. Each of those soldiers is assigned a number. A designated number of soldiers are promoted each month until the list is exhausted during the fiscal year of the announcement.

During the next approximately six months, the new Station Commander did not enlist a single person in the Army. I am serious. He and his family were very nice people without a prejudiced bone in their bodies. We got along extremely well. Our families had frequent visits to one another's homes. The Station Commander would often acknowledge my methods and success in recruiting. He would say things like, "Man, you are something else. I don't see how you continue to make missions each month. Seems like applicants just flock to you."

I would just smile. The fact was, I set my program up so that not only the people I enlisted into the Army, but their parents, spouses, school counselors, and everybody I was in a business relationship with worked for me. I used them as Centers of Influence (COI). In addition, I spoke at several fraternal organizations' meetings. The members of these organizations were very influential members of the community, thus very positive allies for the Recruiting Program. I spoke during High School and College Career Days when they were scheduled in my area. I was an extremely successful recruiter as an individual. However, the station had not made mission consistently in years.

It so happened that six months after the new Station Commander was assigned, the Recruiting Command sent him back to his regular Army job. I don't know exactly who spoke up for me to become the Station Commander after he left, but I suspect it was the First Sergeant because he was the first in my management chain to acknowledge my worth. My opportunity to become the Jonesboro Station Commander came shortly after the previous Station Commander was sent back to his regular Army job. Here's what happened.

As I sat at my desk one day, the Company Commander walked into the station and stated, "Henry Wilson, it's yours." Then he uttered an evil chuckle, with a shit eating grin on his face, and walked out the door.

I thought, *What's with him?* Then I thought, *wait, my Company Commander just made me the Jonesboro Station Commander, but he thinks I will fail.* He didn't offer any type of congratulatory statement or action. All he said was, "It's yours," and walked out the door with that I-got-you-now-sucker grin on his face. Let me give you some background so that you will understand why I thought that.

You see, once I started putting people in the Army, I unintentionally projected a different personality. You know, an I'm-in-control kind of personality, I became very vocal when talking about hiring people to work in the U.S. Army because I had become very knowledgeable of my product (the U.S. Army) and the Recruiting Command. I was, without a doubt, extremely respectful of my prospects, their parents, and each spouse or significant other, and as always, I remained very respectful of my superiors. However, I don't think the Company Commander liked the fact that I would remind him of the new directive banning smoking in government buildings and vehicles.

This was the law enacted by the Federal Government. However, he did not lead by example. Plus, cigarette smoke absolutely made me woozy. He understood and wouldn't smoke, but I could tell he didn't like the fact that I, his subordinate, would remind him, an Army Captain, of the new smoking directives.

That was the law that he was supposed to enforce, and all of his soldiers were supposed to obey. If he had refused to stop smoking in our vehicle and office, I would not have reminded him again. After all, he was the Company Commander. I just think that for me to fail was hopeful revenge on his part. You see, no Station Commander had ever been successful in that station. So, I guess the Commander thought I would fail, also. If my perception was on point, I really surprised him.

Immediately, I started preparing my station for success. You see, now it was about the success of the station, not just me. Jonesboro was now a two-man station with one Army Reserve Recruiter. At this point, I was the only recruiter assigned to the station. However, very shortly, a new Army Recruiter, Gary Dellinger, was assigned. Gary was a Godsend. He was a little crazy (Smile), but nevertheless a Godsend. Prior to his arrival, I had drawn up a plan that I thought would make our station successful. Once Gary arrived, I gave him a complete briefing, during which time I assigned the two northern counties to him, and I took the two southern counties that I had developed, and we would share the home county of Craighead.

In addition, part of the plan was that we would support each other by not enlisting another applicant in our own name until both of us had made mission. As long as both of us continued to work hard to put people in the Army, I would continue my plan. The plan worked extremely well. Both Gary and I were always on the case. Each of us made mission consistently. The entire time we were together we only missed our mission one time, which was a fluke.

I'll explain later. First, let me share this quick story with you. Remember, I told you that my new recruiter was a little crazy, but in a funny way. Gary reported to the station, driving a Ford van with a big, beautiful motorcycle inside. I really don't know how he got that monster in the van in the first place, and, to be honest, I guess I really didn't care. He was a divorcee and alone in Jonesboro, which gave him the opportunity to work late into the evening many times. Now, he knew how I liked a clean and neat office. Check this out.

One morning, I came into the office a little earlier than my normal time, and guess what, I walked into the office and found this humongous motorcycle parked in the middle of the floor. I stood just inside the door, taking several deep breaths to help control my anger before I went about my work. Every now and then, as I sat at my desk, I would glance up at the motorcycle, I guess to make sure I was not seeing things. You know, to be sure, I wasn't going crazy.

Well, a short time later, Gary came running into the office, apologizing as he rushed to remove the vehicle from the office. All I could hear was, "I'm sorry, Henry. I worked really late last night and planned to get here before you and remove this damned thing. Your ass is always on time. I guess that didn't work." Although I was the Station Commander, Gary and I became very good friends, and we worked extremely well together. The way we talked to each other dictated friendship and not a manager-subordinate relationship. During this minor incident, it was so funny, I did all I could to keep from bursting out into laughter.

I looked down at the item I was working on so that he couldn't tell that I was about to crack up in laughter. I just said, "Hey, Sarge, please don't do that again, ok."

He said, "Ok, Henry."

And we moved on. Gary and I stayed in touch for the rest of our careers and even after we both retired. I'll have more about Sarge later in this story. But first, back to my original story.

One Sunday afternoon, I went to my most southern county to pick up an applicant who was scheduled for processing on Monday morning. For reasons I can't explain, a Marine applicant was with my applicant, but he did not have a ticket for transportation to the Military Entrance and Processing Station in Memphis. No problem. Being the person that I am, I took him back to Jonesboro and gave him a bus ticket to Memphis along with my man. Not once did I say anything to the Marine applicant about him joining the U.S. Army. I just wished him luck during his enlistment.

Sometime late Monday morning, I received a call from our career counselor telling me that my applicant had decided to enlist in the United States Marines. Man, was I surprised and pissed at the same time. After gathering all the information, I could by phone, I determined that the Marine recruiter whom I helped by giving his applicant a ticket to Memphis had convinced my man to enlist in the Marines, or so he thought. I was not going to let that happen.

I could not get hold of the First Sergeant, so I called my Company Commander and told him, "Sir, I just wanted to let you know that I am going to the West Memphis Recruiting Station to KICK A MARINE'S ASS. Sir, you may as well start the paperwork now."

I got into my car and headed to the Marine Corps Recruiting Station. Thank God for the hour it took for the trip because it gave me time to cool down, but I went on to the station to confront the Marine.

I was still pissed but under control when I arrived. As I walked into the station, I noticed the secretary through my right peripheral vision as she left her desk. When she saw me, she jumped up from her desk and shot past me like a bullet and ran straight out the door. The Marine jumped from his desk and met me in the middle of the floor, standing kind of in a position of attention. I did not know if he was respecting me or getting ready to kick my ass.

Not hesitating at all, I walked directly up to the Marine to within maybe a foot, stared directly into his eyes with that seriously ugly expression, and stated, "My name is Staff Sergeant Henry C. Wilson. I am assigned to Poinsett and Cross Counties for recruiting. As you work these two counties, I suggest that you ask each person you talk to if Staff Sergeant Wilson is processing them for the U.S. Army. The reason I recommend that you ask that question is if you ever try to steal another applicant from me or my station, you will have my ass to kick." I turned and walked out of the office.

It was kind of funny as I looked back on it. The secretary left that office like turning out a light, and the Marine looked so scared that I did not know what to think. What was really funny is that I could have gotten my ass kicked if I had not calmed down before I entered the Marine Recruiting Station. My message was well received, and neither my coworker nor I had any more problems with the Marines or any other recruiter who worked the same counties as Gary and I.

After leaving the Marine Recruiting Station, I made arrangements for my applicant to meet me in front of the hotel. I picked him up, took him home, and returned to the processing station the next day. Needless to say, I stayed with him until he was processed. As I sat in the Army counselor's office, I received an unexpected order to report to the Military Entrance and Processing Station commander's office.

When I reported to the commander, he stated, "It was reported to me that you disobeyed one of my station regulations."

I asked him, "What was that, sir?"

He stated, "I was told that you recruited an applicant from the hotel we use to house our applicants who are processing for military enlistment. You know that's a violation, Staff Sergeant."

I replied, "No, sir. The regulations do not include the hotel. The regulation states no recruiting applicants from the Military Entrance and Processing Station only."

The commander checked the regulation and dismissed me. Before I left, I explained to him what really happened. I also explained that I would never violate military regulations, nor would I allow any recruiter to steal an applicant from me, because I would never do that to them. All military recruiters are working toward the same goal; therefore, we should not cut each other's throats.

Shortly after our meeting, the commander came to me as I sat in the Army Counselor's Office with an amended copy of the Military Entrance and Processing Station (MEPS) regulation. As I stood and read the amended regulation, I noticed that he had clearly added the hotel to the restriction.

"This copy is especially for you, Staff Sergeant Wilson," he stated. "It appears that you are very well-versed in our MEPS regulations. Thanks for the notification."

"Yes, sir," I said, as I smiled and walked away. Needless to say, my applicant joined the U.S. Army and shipped out within days.

As it turned out, I was very productive during my Jonesboro assignment, just as I had been in every position I worked in since being drafted into the U.S. Army. I have to say, however, that Jonesboro was somewhat different because I learned so much more about people and people's status, from the rich to the poor, as well as people of many different races. My job put me in touch with people from all walks of life. I had contacts with professional people, business people, college and high school educators, laborers, and high school, technical, and college students. I conversed with the rich as well as the poor.

As a matter of fact, when I think about the rich and the poor, the following experiences come to mind. Recruiters are provided with a high school student list for each school within their area, which allows some of the students to take a test provided by the federal government. I experienced a first when I called a student from within Jonesboro, and his mother answered the phone. I introduced myself as Staff Sergeant Henry Wilson from the U.S. Army Recruiting Office in Jonesboro. When I asked if I could speak with her son, the mother's response took me totally by surprise.

"No, you can't speak with my son, and don't call him again." The sound of her voice was unmistakably hostile. She continued, "You need to call all of those poor people out there sitting around doing nothing. They are the ones who need to be protecting us anyway."

I was shocked that she actually made that statement. I was a young man, but not so naive as to think that there were no people who thought they were better than others less fortunate than themselves. However, hearing it so openly expressed took me totally by surprise. I just sat there and pondered her words repeatedly. I would not have believed this story if it had not happened to me. That was a first for me, but yet another learning experience.

Over the course of my life, I came to realize that there are people who have all kinds of prejudices and discriminatory thoughts against others. They dislike those who are not as wealthy, who look different, or simply based on their perceptions of a person or group of people. The sad part is that prejudice will continue as long as it is taught to our children. Let me give you a couple of examples.

After I retired from the U.S. Army, I worked briefly as a manager for a restaurant chain, which included management training. I went through Restaurant Management Training at, from my perspective, a very good restaurant, and became a manager at one of their local restaurants. After I left that company, I worked for twenty-one years for the Federal Government in the Treasury Agency. In both jobs, I experienced some types of racial tension and/or discrimination.

In the restaurant management position, I witnessed racial tension between a black male employee and a white female employee. At different times during my shifts, I heard the black employee using the "N" word openly, where anyone in the kitchen and prep areas could hear him. Since this young man was one of the managers, my intent was to speak with him during a shift when I was Store Manager. However, what I expected came before I got the opportunity to speak with him.

As it turned out, the white employee and the black employee got into some type of argument. As the white employee stormed through the kitchen where employees of different races, including me, were working, she shouted "that damned—" followed by the "N" word. I looked up in shock. I just shook my head in disgust and went on with my work.

A class-action lawsuit was filed, at which time I voiced my concern in writing.

The discrimination I experienced when working for the Federal Government in the Treasury Department was really a surprise. I am talking about the mid-nineties to well into the 2000s, when I was assigned to the Education Group out of Memphis.

During the last several years of my assignment to the Education Branch, I worked out of the Nashville office but was assigned to the Memphis Education Group. The group, managed by a black female, had all females and one male in Memphis and three females and one male in Nashville. I was in my mid-60s and better qualified than any member of the Memphis team. However, each of the females were promoted up to GS-12 ahead of me during a period of just a few years.

One lady, with no managerial experience, was promoted from, I believe, GS-7 to GS-13 so fast that it made me wonder about the manager's intention. What really upset me was when she personally convinced another female who was retiring to stay for a GS-12 promotion that I applied for, and for which I was better qualified. I really could not believe that her superiors let her get away with this gender and age discrimination.

Back to my recruiting experience. As I continued my work throughout my area, I contacted a high school senior who was very interested in joining the U.S. Army. I made arrangements with the school and the applicant's parents to pick him up after school and take him home to meet with his parents. I noticed that the young man didn't have on a coat. At first, I thought it was an "I want to be cute" thing. I asked him in an authoritative voice, "Man, where is your coat? It is too cold out here for you to be without your coat."

He answered, "I don't have one."

Man, did I feel awful! Needless to say, I gave him a coat later. As we got closer to his home, my mind went back to the 50s and 60s as I saw his family standing in the door of their small house. I noticed animals: pigs, dogs, chickens, etc., in the yard, which was fenced in, and an area where there was a garden that obviously had been harvested. I thought, *This is me all over again.*

You see, during the eleven years I was in the Army, I actually lost life's reality from which I came. During those eleven years in the U.S. Army, I was living a good life. Thanks to my U.S. Army employment, I was blessed with three meals a day, nice clothes, and a nice apartment with indoor plumbing. We had nice furniture, and my family was healthy, and they didn't really want for much. I didn't think of the fact that there were still poor people in our country, yes, the United States of America, who still lived as I did before my Army life. As I pulled into the yard of the family home, I handed the young man my coat and insisted that he put it on.

After I met with the parents, their only concern was that their son graduated from high school. That was a no-brainer for me. I convinced them that he would have to graduate from high school to enlist in the Army under this program. As I prepared to end my presentation, the young man attempted to give my coat back to me. Yes, you guessed it. I said, "No, you keep it and think of me each time you put it on." He smiled and thanked me.

My applicant left for the U.S. Army Basic Training shortly after graduating from high school. The family was extremely happy about his anticipated future. This gave me a sense of extreme satisfaction knowing that I was the person who gave a young man the opportunity to be successful in life, just as I had done with many young people during this assignment.

Although I made sure I covered my entire area when prospecting, I put extra special attention into my low-income areas, which consisted of approximately 80 percent of my counties. Contrary to what I thought as I was growing up, I realized that poverty did not limit itself to blacks. It ran rampant through all races of people, especially in my assigned area of northeastern Arkansas.

Although the Black and White races made up the majority of families in my counties, there were some families of other races and some mixed-race families as well. I had the opportunity—yes, opportunity—to visit some of each race of these families. The more I visited and talked with members of families of all races, the more I concluded that, no matter what race, no matter what gender, no matter what religion, or where you live, we are all the same.

We love and, unfortunately, we hate other people for various reasons. What's odd is that when you put life into perspective, no matter what race, religion, or gender you are. It does not matter if you are Gay or what we consider straight; all of us want the same things out of life.

The only difference, as I see it, is that those who were less fortunate want what wealthy people have, and those who are well off or even rich want more of what they already have. I did not let the things I heard and saw stop me from prospecting and visiting anyone who granted me an appointment. I was always professional and treated everyone with respect and fairness. I would enlist a rich prospect into the U.S. Army as quickly as I would a less fortunate prospect.

I didn't care what race, sex, religion, or gender they were. As long as they were not a criminal and were qualified, I put them in the U.S. Army. It was not my job; IT WAS MY PASSION!!! Let me share another experience I had that gave me insight into the mindset of people.

There was a young white female who came into the office seeking employment with the U.S. Army. The young lady lived in SSG Gary Dellinger's assigned area; therefore, he interviewed her and subsequently enlisted her into the Delayed Entry Program (DEP) with a departure date not too long in the future. As it turned out, this young lady was another great experience for me. She visited the office regularly during the period prior to departing for active duty.

Strangely, however, each time the young lady visited, she would stop by and talk with my friend Gary for a short visit. Since Gary was her recruiter of record, I expected her to spend most, if not all, of her time in our office with him. After her short visit with SSG Dellinger, she would come directly to my desk. The odd thing was that she would sit and talk with me for as long as I would allow her. I am kind of a direct person if I need to be, so after she visited the office several times and sat with me longer than she would with her recruiter, I felt obligated to find out why.

This time, after she completed her usual routine and came to my office, we talked for a few minutes. This time, I came straight to the point. I asked her why she would spend more time with me than with her recruiter of record. What she told me was kind of shocking, but not really. This time, it was really good for me to erase the perception.

The young lady told me, "All my life, my parents, especially my mother, told me not to mix with Negroes (Black people). They are very nasty people. You can catch all types of diseases from them. She never told me what diseases I would catch, just that I would catch diseases."

The young lady went on to say that when Black people would come into their drugstore and buy things, her mother would never take the money directly from them. She would have the Black people lay the money on the counter, and her mother would lay the return change on the counter for them to pick up. The young lady went on to say, "My mother told me that she never touched a Black person." She said, "So, my mother's actions were so adamant that I became extremely curious. The young lady continued. I never told my parents this, but I started talking with different Blacks and mixing with other races of people to see if there was any truth in my mother's beliefs and perceptions. I only started doing what I am doing now after high school because I did not have many opportunities to socialize with Blacks. After high school, I started college, and that's where I began socializing with people of different races."

The young lady told me that some of her friends she met in college, White and Black, were joining the Army for the financial and educational benefits. She told me that she really didn't need the money, but joining the Army sounded exciting to her. She said that when she saw how well my coworker and I worked together, she became more excited about joining the Army. She told me, "When I saw you, I thought this was another opportunity to learn more about Black people."

I told the young recruit, "I think what you are doing is a good thing. If more people would not act on what they hear or what they believe about a person, a race of people, a people's faith, or sexual orientation, I think there would be a lot less hatred in the world."

I cautioned her, however, not to force herself on people. "Just go about your work or play or whatever you are doing and be aware of your surroundings. Pay close attention to people of any race. I think you will find some good and bad in any race of people. I think you will find that we are all the same inside, just a little different on the outside. Don't be gullible. Be vigilant of all the people around you. You will be surprised by what you will see in people."

I let the young woman know that I was very impressed with how she approached her curiosity and that I thought she would do well during her tour of duty in the U.S. Army. I told her to always do her job, but understand that not everybody who comes into her life has her best interest as their priority. Neither do they intend to hurt her in any way. This was one of my most rewarding experiences.

CHAPTER 21

Community Service

My wife and I attended a small local church of which we had the honor of serving as honorary members on its Board of Directors. Why honorary? Well, we really didn't believe in the way that denomination managed its finances. Let me explain.

We did not like the fact that the church was required to send a dictated amount of funds to the managing higher church each year. The real problem was that once the local church sent money to the higher church, more often than not, it left them without enough funds to take care of their own business. For example, we sat in church and watched it literally rain through the ceiling. However, they could not get funds from the bishop for building repairs or anything else. We actually informed the members that we would never become official members of the church; however, we would support their local needs.

This is where Sarah and I came in. The church was in such bad condition that I thought repairs would be a waste of money. We convinced the church leaders and members that everyone would be better served if they built a new church. My wife and I were instrumental in helping the members raise the first five hundred dollars to build a much-needed new church. I headed up a committee to raise funds to build the church.

I made sure that the stipulations were that the money raised was to be deposited in a Building Fund Account. The funds would not be a part of the funds raised during each church service. We stipulated that the funds would be used for building a new church only. My wife and I also directed that our offerings each Sunday would be added to the building fund for a new church. The Church Building Fund was overwhelmingly approved by the members.

Our youngest daughter started school in Jonesboro as our oldest daughter continued her education there, where she attended Annie Camp Middle School. Arkansas State University was a major college in northeastern Arkansas. Several of our associates and friends graduated from Jonesboro High School and went on to graduate from Arkansas State University. One of them, whom I had the opportunity to meet, went on to become Miss America in 1990.

I also had the opportunity to meet and socialize with many great people who went on to achieve great success in their fields. Our best friend, Deborah Agnew, recently retired as a School Counselor for the Jonesboro Public School System. Deborah received her Bachelor of Science in Education, Master of Science in Education and Master of Science in Counseling from Arkansas State University within 16 years. Although receiving numerous awards during her career, Deborah's greatest achievement was that she established a pilot program in partnership with ASU for a Jonesboro school to become the first elementary school in the state of Arkansas to provide mental health and social services programs within the school.

Deborah's biography is a book within itself, so I will leave it at that. Another wonderful friend, Jane Gates went on to become a college President, another competed in the High Jump during the Olympics, and another went on to coach college football, just to highlight a few success stories. It was an honor to socialize with those great people.

Several years after I transferred from Jonesboro to Texas, my wife and I had the opportunity to return to Jonesboro on business. Of course, we took the opportunity to visit our friends. We also had the honor to attend Sunday service in a brand-new church during its anniversary. The church was small and beautiful, but very appropriate for the membership. The guest speaker happened to be the pastor who was in charge when we left. He had since been assigned to another church. During the pastor's opening remarks, he introduced us and thanked us for being the catalyst that pushed to start the building fund that helped to build their church. We felt great about the success of the members.

CHAPTER 22

Recruiting Awards
and Progress

I continued to flourish in my recruiting duties. Receiving an award, in most cases, was a surprise to me, but they kept coming. You see, I never worked for awards; therefore, I never spent time counting the required points to earn them. Most of my counterparts knew how many points they needed to earn their next award. Don't get me wrong, I am not faulting them, but for me, that was time lost from recruiting qualified applicants for the U.S. Army. I'll have more about my awards later.

The Army opened enlistment for females to enlist with the same test scores as men in approximately mid-September of 1977 or 1978. Previously, females with a high school diploma had to score much higher on the entrance exam than male high school graduates to enlist in the U.S. Army. The effective date that we could enlist females under the new requirements was about two months after the beginning of the fiscal year. Talking about being proactive. I, just as many recruiters throughout the U.S. Army, had so many females in QNE (Qualified but Not Enlisted) status before the effective date that it prompted the Recruiting Command to order the effective date moved forward.

I had enough females prequalified to make mission for the next two months. No, I didn't hold them. I didn't believe in delaying an enlistment to make mission. If they were qualified to enlist, I put them in the U.S. Army on the very first day possible. My job was putting people in the Army. Sure, I knew the award system was there, but my motivation was success. My attitude was, do your job, and everything you deserve will fall into place.

I would be remiss if I didn't share another bit of experience I gained while serving in Jonesboro, AR. One day, as I sat at my desk taking care of Recruiting Station Administration, a father brought his teenage son into our station to enlist. I greeted them and introduced myself. The father, who introduced himself as a Baptist Minister, did all the talking. He was very pleasant but talked a lot.

As it turned out, he wanted his son to join the Army, but he only wanted him to work in the Food Service Career Field. I was flattered because, as you know, food service was my first job, and I loved it dearly. However, my office policy was that each recruiter would give our prospects an overview of the Army and the job or jobs they qualified for once they took the mental and physical tests.

The mental and physical tests are what dictate which career field each applicant can be enlisted in. The father agreed to allow his son to take the mental and physical tests and enlist in the Army if qualified. Of course, the young man agreed and seemed comfortable with the process, and, most of all, he was ready to enlist. I completed the application and took the young man to the MEPS in Memphis for processing. He passed both the mental and physical tests with flying colors. In the written test, his son qualified in the highest category. Consequently, the Career Counselor was required to give the applicant an overview of all the jobs that were available for him to choose from. You probably can guess that there were many jobs available for highly qualified soldiers.

The U.S. Army also gave qualified soldiers a great start financially and educationally. The young man signed a contract for a job nowhere close to food service. He and I thought his father would be extremely happy. However, when the young man informed his father of his contract, to his and my surprise, the minister blew a fuse.

He called me, yelling at the top of his voice. He accused me of lying to him and tricking his son into signing a contract he didn't want. I tried to get a word in so that I could explain to him what happened. This is how the conversation went:

"Sir, sir, please let me explain what was done." He kept yelling.

"Sir, sir, if you would calm down, I will explain what happened."

Continuing to yell, his response to me was, "You don't talk to me like you talk to all those other misfits in the army."

Now I am pissed, but I kind of kept my cool. I responded to the preacher, "Sir, that was an awful thing to say." Then I hung up the phone.

He called right back. "Sergeant Wilson, I was not talking about you."

I interrupted him and stated, "Sir, I am with the United States Army in Jonesboro," and I hung up again. He called back again and was clearly embarrassed by his voice and his actions. The minister calmed down and listened to what I had to say.

I explained our procedure until I was convinced the father understood the process in detail. However, most of all, the son convinced his dad that this contract was what he wanted. I asked the father to take a few days and to please think about the advantages the contract would give his son. I explained to him that, afterward, if he still wanted his son to enlist for the Food Service job, I would have his contract changed to the Food Service field.

Evidently, the young man convinced his dad that this was his choice, and the job he chose was what he wanted to do. The son also told me that he explained to his dad that he would receive a cash bonus for enlisting for the job. The recruit kept his contract and shipped out not long afterward. I believe the issue was that the father had no confidence in his son's abilities.

Here's what I think is the best part of the story. Check this out: Several months later, the father surprised me with a phone call. He told me that his son was doing really well in the U.S. Army. The other reason he called me was that his brother, who lived in Little Rock, AR, wanted me to put his son in the Army. The preacher said that he was bragging about me and how hard I worked to get his son what he wanted.

Of course, I was really flattered. I made arrangements for his brother to talk with a recruiter in the Little Rock area.

I really felt good about this entire experience because the minister's action meant that without a doubt, I caused him to change his perspective on the U.S. Army. And he is, no doubt, a great Center of Influence for our recruiting efforts. What's even better is, I believe that there is nothing greater than a parent's trust and support for their child or children's desires.

The one negative situation I experienced during this assignment was when I was unexpectedly suspended because of an alleged malpractice. Let me explain. I, along with several other recruiters from within the battalion, was ordered to report to the Battalion Commander in Jackson, MS. While there, the Battalion Commander went over the malpractice charges for each recruiter by name. I was the last recruiter he covered. The Battalion Commander simply stated, "Staff Sergeant Wilson, I have no idea why you are on the list."

However, when I received the packet from USAREC HQ, it stated that the Company Commander of one of the soldiers I recruited stated that, and I quote. "The soldier had to have help to enlist in the U.S. Army." When I saw the soldier's name, I remembered him right away. The Company Commander actually stereotyped the young soldier because of his accent. Keep reading.

After I read the allegation, I really could not believe it. As I said, I remembered the young man in question. He was a young white man around 21 years of age with an unusually deep southern accent. When I conducted my initial appointment with this young man, I noticed that he spoke proper English, but he spoke unusually slowly. It seemed almost like he was unsure of what he wanted to say or was afraid to say. All I can say is that his accent was different enough that you really had to listen attentively in order to catch the meaning of what he was saying. During my prequalification of the applicant, I realized that I had to let him talk without interruptions. He was impressive during prequalification. Therefore, I scheduled him for the mental and physical examinations. The young man passed both tests, so I put him in the U.S. Army.

One of my qualities is that I don't discriminate against a person because of their dialect, accent, or mannerisms, because they are different from what some people call normal. I believe the young man's Company Commander did just the opposite. Check this out.

When I received the packet with detailed information that was used to suspend me, I was totally shocked. The young man's Company Commander's statement was something like this: the Company Commander said that when the soldier talks, you can't understand him. He had to have help to enlist in the Army. My advice to anyone who finds themselves in a situation like this one is to try listening better and try confirming what you THINK YOU heard by repeating it to the speaker. Be patient and do not dismiss the person as being dumb or not as smart as you think you are.

Right away, I prepared a rebuttal to my suspension. The suspension was overturned by the USAREC Commander. However, the Brigade Commander forwarded the packet to me with an addendum stating that the approving authority had cleared me. The Brigade Commander went on to state that he wanted me out of his Brigade. He directed me to request reassignment immediately. I could not understand why the

Brigade Commander was so hostile toward me, especially since the General had cleared me of any wrongdoing. Being the soldier that I was, I followed the Brigade Commander's orders.

I completed my reassignment request and sent it forward just as I was directed. When I received the response from the Brigade Commander, I was still working out of my Company Headquarters. The Company Commander received the response, read it, and handed it to me. The response stated that I had three reassignment choices with the locations listed. He went on to say, "Or I could withdraw my request and stay where I am. After I read the directive and laid it to the side, my Company Commander stated that he didn't understand why the Brigade Commander abruptly changed his mind.

I simply stated, "I understand clearly, sir," and continued doing what I was doing. I did not offer an explanation to my Company Commander; however, I believe that the Brigade Commander superseded the General's orders when he ordered me out of his Brigade.

When I submitted the request, I wanted it to be as completely correct as possible. Not thinking that it would have any effect on the Brigade Commander's decision, I used the subject, "Per Order of the Brigade Commander, I request reassignment from his Brigade." I think that's what caused him to change his decision.

My coworker and I continued to be very successful in doing our jobs. Our Jonesboro Recruiting Station was on the radar within a large portion of the Recruiting Commands. As a matter of fact, my coworker, Gary Dellinger, was promoted to the grade of Sergeant First Class and immediately transferred to a recruiting station in Dyersburg, TN, as Station Commander. I totally agreed with the promotion and the assignment, but man, did I hate to see him go.

We worked well together and had become very good friends. His last day in the office was hard for me. I remember sitting at my desk as he walked toward the door. He paused for a moment as if he was waiting for me to say or do something. I sat, and for fear of crying, I didn't say a word. I glanced toward him, threw my left hand up, and muttered a weak goodbye.

I was too emotional inside, and I never got up from my seat to shake his hand or hug him, as real men do, for fear of doing what I really felt I would do, cry like a baby. So, I just sat there looking down at my desk and just waved goodbye. We kept in touch for years and had the opportunity to work together again. Keep reading.

Another recruiter was assigned to me shortly after my friend's reassignment. Yes, he was white, which I expected because I knew that my superiors wanted a Salt-N-Pepper Team in Jonesboro. I had no problem with the race of any soldier assigned to me. My job was to train new soldiers to be the best recruiters they could be. I learned early on in my life that race has nothing to do with a person's personality or their learning ability.

To this day, I know that hating someone because of the color of their skin has no place in my heart or mind. My new recruiter was a very sharp Noncommissioned Officer. We worked together until I transferred to our U.S. Army Counseling Office in Memphis. I also had an Army Reserve

Recruiter assigned to me around the same time. She wasn't there very long. I really didn't get to know her very well.

As I mentioned earlier, I do not work for awards. I always believed in doing my assigned job to the best of my ability and then some. Needless to say, my awards kept coming. Don't get me wrong, this did not keep me from checking the promotion list when it was posted each year. While analyzing the lists, I couldn't help but notice that a number of my counterparts who were promoted were, or had been, career counselors during some period in their careers. No problem. I programmed that info into my memory bank and kept doing my job. Keep reading.

I realized in my later years of life that God never leaves your side. However, sometimes He does things for you, and you wonder to yourself, "Why did that happen at this precise time?" That's how God does. You just have to know when you are blessed. As they say, "You better recognize." (Smile.) One example of this is my Company Commander informed me that the Battalion Commander was visiting the Company and would be stopping by my office for a visit and to present me with yet another award. My first thought was, "Award for what?" But I just responded, "Thank you, sir."

After the Battalion Commander completed his awards presentation, which included telling me how he appreciated the great work I was doing, I thanked him, and at that point, I took the opportunity to ask the Battalion Commander if I could have a few words with him. We went back and took a seat in my office. I simply told him that I loved field recruiting but needed to be a Career Counselor. I went on to explain,

"Sir, based on my analysis of the last two promotion lists, most of my counterparts who received promotions were career counselors or had career counseling in their background. Sir, I would like to become a Career Counselor."

The Battalion Commander simply stated, "Ok, Henry, I will look into that."

The Battalion Commander kept his word and called me the very next day with great news. He stated, "Henry, I have directed my staff to get you into the Career Counselor Course right away. I found out during this situation that if you had not shared your concerns with me, you probably would have been left behind."

Of course, I was confused until he continued. He said, "I was told with complete seriousness that there are leaders within your management chain who didn't want to move you. Looking at your record prompted me to ask why." The Battalion Commander continued, "I could not believe what they told me. I was told that you are the only Station Commander who has ever made the Jonesboro station successful." He said, "Henry, while this statement is a compliment to you, it is also discriminatory toward a great soldier. Thanks to you, this will never happen again, not on my watch anyway. I am really glad we talked. Thank you, and good luck in training."

I responded, "Thanks to you, sir. I really appreciate what you've done for me." I hung up the phone, smiled, and thought, God did it again. Within a few days of our conversation, I received the notice that I had a slot in the very next Career Counseling class.

CHAPTER 23

Career Counseling Course Preparation

My First Sergeant came into my office a few days after I was notified of my assignment to inform me of my class start date, which was March 6th, and walked out the door. Before the door closed, he walked back in and stated, "You know you must be within the Army's weight requirements when you get to school, don't you? If you don't meet the weight standards, they will send you home." He laughed as he walked out the door.

Check this out: he walked back in, gave me the class starting dates again, and walked out. As he walked out, he actually gave that same chuckle again, but with a more pronounced "I got you now" laugh. I thought, *What's with him?* Then reality hit me right in the face. I thought to myself, *Henry, you are overweight.*

I went to the back and weighed myself. Man, was I shocked. I was 25 pounds over the U.S. Army's weight standard for my age and height. The class started in three weeks—yes, three weeks. I thought, *What am I going to do?*

At that point, I realized my supervisor laughed because he knew I wouldn't lose the weight. Therefore, I would be sent back as the Station Commander. As my supervisor, he was supposed to help me set up a plan to lose weight. But no, he was more interested in having me return as Station Commander than in helping with my weight problem. As a supervisor, that was an attitude I would never tolerate in myself. You see, I understand that the soldiers assigned to me are my most valuable assets. No problem, though, I went for myself. I devised a plan to get within the weight limit before reporting to the course.

How did I do it? First, I cut my food intake to eleven hundred calories per day, and I only ate broiled or baked foods, and very little of that. I jogged and rode my bike several times a day, religiously. In addition, I sat in the sauna of a local spa at least once a day. Man, was it tough. The worst part happened during the third week. Not only was it our tenth wedding anniversary, but to celebrate, my wife and I had also planned a weekend in Memphis, which covered the 3rd of March, our anniversary day. At that point, no matter how hard I worked, the pounds stopped dropping. Man, was I depressed, but I stayed the course.

Understanding that I had to report to class on Monday, March 6th, and that I had three pounds to lose in three days before we left for the weekend, I thought it was best to cancel our plans. Reluctantly, I called my wife and explained my dilemma. About 30 minutes later, she walked into my office, laid an envelope on my desk, and turned and walked out without saying a word. Although she did not say anything, her actions spoke louder than anything she could have said. I opened the envelope and realized that she had made hotel and restaurant reservations for us to use during our celebration.

I need each of you to understand something. To this day, I have always put my family first. My wife's feelings were extremely important to me. I dropped my head and thought to myself, Oh God, what do I do now? First, I called my wife and apologized, and told her that we would not change our plans. For the next three days before we left for Memphis, I continued working hard to lose those three pounds.

My wife and I spent the weekend in Memphis and had an amazing time. I managed to put my personal problem behind me, at least during our wedding anniversary celebration. Since the Memphis area is our home, we purposely did not visit anyone. As a matter of fact, we ran into one of our high school teachers who apparently worked at our chosen restaurant. We spoke to him briefly and kept stepping. (Smile.) We intended for this celebration to be just the two of us so we could do what we wanted, when we wanted to do it, and we certainly did!!

Come Sunday morning, as we prepared to leave Memphis, I couldn't help but fantasize about our next anniversary. My wife departed for home, and I departed for the Counselor's School in Indianapolis. Leaving Memphis was not pleasant because I was extremely nervous about my weight. However, it was necessary for my career. Before I departed, I stopped by a local grocery store—yes, you guessed it—and picked up a box of Ex-Lax. I sent two squares down the hatch and hit the road.

I arrived in Indy a few hours later. I took two more squares of Ex-Lax and went to bed for the night.

Early Monday morning, on March 6, 1978, I prepared myself for the worst. I was extremely nervous when I reported to the training room. The lead instructor lined all students up in alphabetical order and began the weigh-in process via the student list. Because I was the last student in line—Wilson, yes, you guessed it—I was more nervous than for any position change I had ever made.

Here we go. I stepped on the scales, with my uniform hanging off me, and stood there for what seemed like forever. Then the instructor said, "Two hundred and three pounds."

I shouted, "Thank you, Jesus," as I stepped off the scales. If the truth were known, I was so nervous the scale needle kept jumping between two hundred and two and two hundred and four, so the instructor just took the middle. By the grace of God, I made it. Hold on, I'm not through yet. Check this out.

The course was very simple. I had no problems with any part of the training except the computer program that I had to use to secure contracts for each applicant. The problem was that each step had to be completed in sequence. The instructor told us that we could not miss any step. What was even worse was that this phase was one-third of our grade. There were three phases. Yes, you guessed it. If you failed this phase, you fail the course, because 70 percent was the minimum acceptable score. This was the last test of the course, and on the last day, the first thing in the morning.

What I should tell you is that I don't take tests well. You can't beat me on the job, but tests simply stress me out. As I sat in the classroom, just staring at my computer, it was freaking me out. Therefore, I just closed my eyes and sent up a quick prayer. Just about that time, the lead instructor came into the classroom and made the most profound announcement I have ever heard. He stated, "Morning, students. I have bad news and good news. The bad news is we are very sorry, but the computer system is down, and it is going to take a few days to get it up and running. The good news is, because of this glitch, we must give you all 100 percent for this phase of training."

Every student shouted and cheered. I just looked up toward the heavens above and silently said, "Again, I thank you, Jesus." I was so relieved that I could feel my body as it began to return to normal.

CHAPTER 24

Counseling Duties at the Military Entrance and Processing Station (MEPS)

I returned to my Jonesboro station, packed my things, and said goodbye to my coworkers. The fact that I lost 25 pounds in three weeks obviously shocked my Company Commander and First Sergeant. This is simply another reason we should never assume what another person won't or can't do. Both my Company Commander and First Sergeant knew I was overweight, but they never knew how much, simply because they never weighed me. Still, I believe they never thought I would make the minimum weight in time to attend the course. Anyway, I reported to my Career Counselor position at MEPS the following week after school.

Man, was I a natural! My goal was to secure a contract for each applicant I processed for enlistment into the U.S. Army, and that's what I did. I never sent a mentally and physically qualified applicant home without a contract. You remember the computer system the MEPS used to secure contracts for applicants? Yes, the one that went down when I was in training. Well, as it turned out, the system was not as smart as I thought, or I was smarter than I thought. (Smile.) Allow me to share this one story with you, and I will move on.

I took pride in securing a contract with each applicant I processed, no matter their race or sex. I had already done an analysis of the entrance test from the study practice test that was available for the applicants. I concluded that the vocabulary and paragraph comprehension portions of the test were biased. Why, you ask? Well, looking at the words and situations listed in the paragraph comprehension portion of the test, depending on your surroundings as you grew up, you may never have heard of some of the words. Therefore, how could you use them when communicating with others, or how could you even spell them or use them in a sentence?

During the year I worked as a counselor, I realized that the system was programmed according to the needs of the U.S. Army. I concluded this because, when we built a file in the system, we had to add each applicant's race and gender, along with their mental and physical qualifications. Consequently, the system did not show all of the jobs that minority or female applicants—or those male applicants who did not graduate from high school—qualified for. Nor did the system show the same jobs for Category Cat-4, Cat-3, Cat-2, and Cat-1 applicants unless they were needed in those jobs. The Cat-1 covers the higher mental scores, and Cat-4 covers the lower mental scores. The number of job availabilities could change periodically during the day.

Note: higher scores on the mental test for an applicant dictate the number of job availabilities for that applicant. I believed that when I requested a job for a Cat 4 applicant, the system would show fewer jobs than it would for a Cat 1 applicant. The system would only show jobs that would allow certain types and categories of applicants to enlist. I always recommended that an applicant have his or her recruiter send them back as many tries as allowed.

I had studied the system enough to realize that no system is foolproof. As it turned out, I got my chance to test my theory on two females who happened to be Black. The other counselors had processed the two applicants but could not secure a contract. After we finished processing all of the other applicants, I quickly went and retrieved the two applicants' packets from the completed file. I tried my theory in several different ways on the system.

Finally, I followed the required sequence to secure a contract over and over several times until it worked for the first applicant. Therefore, I quickly tried it with the other applicant, and it worked. I called each of the applicants, one at a time, and processed them. They were extremely happy.

When I took the first contract to the Senior Counselor to complete the process, the other two counselors came running to my desk to find out how I secured the contracts. I just smiled and said that I was just trying repeatedly. I made sure the contracts had all the verification information required on each one. I loved hiring people for the U.S. Army, but the two applicants were special to me. It was obvious they wanted to join the U.S. Army because they had sat there all day. Each time

I glanced over at them as I continued my work, they looked so dejected. I would just smile, trying to give them a little hope. I also reassured them that I would take a look at their packets shortly. The fact that I was able to secure a contract for those two young ladies to enlist in the U.S. Army made me feel really great.

The two applicants came by my desk to thank me once they were sworn into the U.S. Army. I told them to just do great things as long as they were soldiers. Each of them replied, "Yes, sir," and went on their way. I was extremely happy that I was able to secure contracts for a bunch of applicants during that year I worked as a counselor in the Memphis AFEES.

CHAPTER 25

My Promotion to Master Sergeant

The conversation I had with my Battalion Commander worked. Exactly one year after I started working as a Career Counselor, I was selected for promotion to the rank of Master Sergeant (E8). Until this promotion, I did not realize that there were so many people looking for me to work in their area of recruiting. As soon as the list was posted, my Battalion Sergeant Major told me via a phone call that I would become the First Sergeant of the Jonesboro Recruiting Company as soon as the assignment was approved.

Just to refresh your memory, I was assigned to the Jonesboro Company as a recruiter and later as the Station Commander. Then I was assigned as a Career Counselor in Memphis. As I mentioned earlier, the Company Headquarters was located in Jonesboro, AR. Here is the great part. I never moved my family from Jonesboro. As a matter of fact, my wife's office was in the Federal Building on the same floor as my would-be office, literally within fifty to seventy steps. This was really too good to be true. Anyway, I called my wife, Sarah, right away and gave her the good news. Man, was she excited— even more than I was. Well, that joy lasted about as long as a snowball in hell. Well, just a little longer than that. (Smile.)

As I sat in my work area doing what I do, I got this phone call. Our clerk answered the phone, "SFC Wilson, phone call for you." I answered, "This is SFC Wilson. Can I help you?"

"Is this SFC promotable, Henry Wilson?"

I responded, "Yes, sir."

He continued, "This is Sergeant Major Mata, Battalion Sergeant Major of the Oklahoma City Recruiting Battalion."

I responded, "Hi, Sergeant Major, what can I do for you?"

The Sergeant Major said, "Are you ready to come this way?"

I snickered and said, "I don't understand, Sergeant Major."

"Are you ready to become my Senior Counselor in Amarillo, TX?"

"Oh, Sergeant Major, I understand now. I'm sorry, but I am going to be the First Sergeant of the Jonesboro Recruiting Company."

"No, Hendry"—that's the way he pronounced my name— "Hendry, you are coming this way."

I paused for a moment, wondering what was going on. Then I said, "Do you mind if I call my Battalion Sergeant Major?"

The Sergeant Major said, "No problem. Call me back."

I called my Sergeant Major. "Sergeant Major, this is SFCP (the P is for promotable) Henry Wilson. I just got a call from the Battalion Sergeant Major from Oklahoma City telling me that I am being assigned to Amarillo, TX MEPS as the U.S. Army's Senior Counselor."

My Sergeant Major responded, "Yes, Henry, we lost the battle. Higher HQ got involved, and we had to let you go. They really need a person of your caliber to fix things in that shop out there. We really hate to lose you, but I am confident that you will do a great job out there."

You know, I never stopped to think about what the Sergeant Major was saying about me until many years later. I mean, I always reflected on my military career and the positions I held in many sections in different career fields because I really enjoyed my career. It was many years later, after I had retired from both the Army (after 24 years) and my second career with the Federal Government (after 21 years), that I thought to myself, "Man, I must have been something."

I had high-ranking officials negotiating for my service. You see, all I ever did was try to do the best job I could in whatever position I was assigned.

Anyway, back to Amarillo. I called my wife, Sarah, and gave her the news. She checked right away and found out that there was no position in Amarillo, TX for her. No problem. The government agency she worked for created a position for her. You see, she was a super employee also (Smile.). This really helped to alleviate the stress of relocating my family from Jonesboro, AR, to Amarillo, TX.

CHAPTER 26

Senior Career Counselor, Amarillo, Texas

Prior to my reporting date, my wife and I made a trip to Amarillo to find an affordable place for me to live. Why just for me? Well, since my reassignment took place early in the year, my wife and I agreed that we would never have our children transfer from one school to another during a school year.

We didn't want that added burden on our two daughters. We found a small efficiency apartment, which was great for me. My wife took the opportunity to visit her future office. The employees there were very accommodating. She looked forward to working there. Anyway, we returned to Jonesboro, where I prepared to report to Amarillo for my first day of duty. In just a few days, I departed for Amarillo, looking forward to my first Senior Counselor position.

As I traveled to Amarillo, I made a mandatory stop at the Recruiting Battalion Headquarters in Oklahoma City for in-processing and the initial briefing by my immediate supervisors, the Battalion Commander, Lt. Colonel Davis, and Battalion Sergeant Major Mata.

Most of you may remember the Oklahoma City bombing. Well, the Battalion Headquarters was in that Federal Building. Man, did I cringe when I heard about that bombing, because I had been in that building many times. That's where most of our staff meetings were conducted. Although I had retired from the U.S. Army when the bombing occurred, knowing where the perpetrator parked the truck that housed the bomb really sent shock waves through me.

I mean, having been in and out of that entrance many times made me stop and think that the bombing could have happened during one of my visits. Thank God it didn't. May God be with the dead and bless those who were injured and their families.

My meeting with the Battalion Leadership was very pleasant and informative. They covered my responsibilities and the office layout. The information that stuck with me most was when the Sergeant Major emphasized, "Hendry, that office is yours. The Military Entrance and Processing Station (MEPS) is within this Battalion's area of responsibility." He continued, "There are two E8s working out of that office that outrank you. However, you are in charge of that office. You are responsible for everything that happens or fails to happen in that office."

My response was, "Ok, Sergeant Major. Not a problem."

That was always my response anytime I was assigned to a different position. Even after I retired from the U.S. Army and went to work with the Federal Government, I always responded with an affirmative. Anyway, the one thing that they failed to tell me really was a shock when I checked into my new office. Keep reading.

During mid-afternoon, on, I believe, a Tuesday, I went directly to my office, which was inside the Amarillo MEPS. I walked into the office, where there were several soldiers sitting at two separate desks, to introduce myself. As I conducted a quick scan of the room, I noticed a Master Sergeant, a Sergeant First Class, and a couple of civilians, whom I assumed were applicants processing for enlistment.

"I am SFC Promotable Henry Wilson, the new Senior Counselor for this office," I stated. Here is what really surprised me. The Sergeant First Class sitting directly inside the entrance of the office, obviously surprised, stated boldly, "I am the Senior Counselor of this station!"

As I stood there in surprise, I couldn't help but notice the oversized ashtray filled with cigarette butts, each of which was standing on the filter end. They formed circles from the outer edge of the ashtray to the center. The circles were perfectly formed. In my opinion, the action was very nonproductive.

I hesitated for just a moment and then responded, "I hear what you are saying, Sergeant First Class, but you probably ought to call the Battalion Sergeant Major." I was really shocked that the Battalion Sergeant Major did not tell the soldier that he was being replaced. All I know is I never saw him again. Not my problem. I moved forward aggressively with what I had to do.

During a briefing by the Master Sergeant, who was the counselor assigned to the Albuquerque Battalion, I learned that the Sergeant I replaced wasn't very customer service-oriented. I was told that he was in possession of many deficiencies. None of that information had anything to do with what I had to do. As I said, it's not my problem. However, it did give me insight into why management wanted him out.

I was really surprised by the layout of the office. Let me explain. The office consisted of three nice-sized rooms. One office was used by a Master Sergeant Reserve Recruiter/ Counselor who was there maybe two days a week. One room was shared by the Albuquerque and Oklahoma City counselors. The other room was an overflow room.

The counselor from the Albuquerque Battalion had a computer unique to his battalion. I also maintained a computer unique to my battalion. The difference was that I was in charge of the site, which was made clear by my Battalion Sergeant Major. However, both counselors shared the computer that we used to secure applicant contracts. After studying the layout of the office and considering what we had to do, together with applicants and other traffic, I proceeded with making several changes to the office layout right away.

As soon as the Reserve Counselor/Recruiter came into the office, I took my first opportunity to meet with him. I informed him that he would use the overflow room when he needed it. I separated myself and my co-counselor so that we had our own office to process applicants assigned to our individual battalions. I also moved the computer we used for securing contracts into my office, placing it where my co-counselor had complete access to process his applicants. Moving forward with our work proved that my plans and changes worked perfectly.

During the process of getting my office in order, I took my first opportunity to introduce myself to the MEPS staff and learned the layout of the MEPS as a whole. I also needed to know who was in charge of each section within the processing center. This was extremely important because I needed to know who to go to for whatever I needed to get the job done. Filling the U.S. Army's ranks with qualified people was the name of the game, and I intended to make it happen.

Gaining the cooperation and respect of the MEPS staff was the first step in getting to where I needed to be. Within a few days, yes days, I convinced my coworker that, although we were working for separate battalions, we had the same mission. There was no reason why we couldn't work together. I went on to say that I was willing to help process his applicants when I was not busy with mine. He agreed that he would do the same.

As it turned out, we worked with the applicants as they came in, no matter what battalion they were assigned to. Although my coworker did not work under my supervision, this plan was similar to the plan I set up in Jonesboro with my friend, Gary Dellinger. The only difference was that Gary and I were assigned to the same unit, and in Amarillo, Master Sergeant Wesley was assigned to the Albuquerque Battalion, working out of my station but not under my supervision.

He enlisted applicants out of northwest Texas and eastern New Mexico. Once he agreed to my recommendation, we were off and running. We made sure every qualified applicant was enlisted in the U.S. Army.

I've got to tell y'all, I had fun in this position! Although my counterpart and I agreed to the new plan, each of us was totally responsible for our individual battalion. For one year, I was totally responsible for getting qualified applicants for the Amarillo Recruiting Company to sign a contract for the U.S. Army. I did extremely well.

One counterproductive thing I noticed after my assignment was that some of the Amarillo recruiters had developed a trend of prequalifying a large percentage of their applicants. They would then process the applicant for enlistment if they passed both tests. Let me explain prequalification. A legitimate prequalification is needed when an applicant has a disqualification that needs the commander of a higher headquarters' approval to enlist. The problem was that there were recruiters who, for some reason, stereotyped the applicant. They didn't think the applicant could pass either the written test or the physical. I'll explain.

Their belief, especially on the mental test, was based totally on a stereotype they formed during their interaction with the applicant and/or their family, but especially their perceptions from the past. The recruiter could have based their belief on race, dialect, their living environment, or any other unfounded reason. Prequalifying applicants for any reason other than for Company Commander or higher command approval to enlist the applicant is counterproductive and unacceptable to me. Stereotypes, for whatever reason, are a practice I would never condone. The fact that I was not in the recruiter's supervisory chain and that I had my individual job to do didn't allow me to go into a training mode with each individual case. I'll explain what I did later. Keep reading.

First, I must share some stereotypes I have witnessed during my lifetime. I have already mentioned a few of them; however, my experience with stereotypes goes all the way back to high school. There were a few teachers who treated some of the students differently because of their parents' status in the community or how successful they thought the student would be—or not be—in life based on their family background. Sometimes the student's scholastic ability played no role in the teacher's decision. In other cases, their perceptions could have been based solely on hearsay or, in some cases, the actions of someone's siblings.

My father-in-law once told me that when the word got out that his daughter and I were getting married, a friend of his, an old man, made a serious point of telling him, "Jim, you ought not let your daughter marry that boy. That boy ain't no good."

I knew of the man, but I had never said a word to him, and he had never said a word to me. Hearsay is what the old man based his opinion on. The problem is that the old man acted on hearsay. However, my father-in-law responded, "My daughter is marrying Henry. I'm not. If things don't work out between them, my daughter knows that she can always come back home."

Over fifty-seven years later, Sarah and I are still married and are doing outstandingly well while continuing to do our thing together. We have three very successful adult children and six grandchildren. The fact is, those mouths that speak without justification should remain shut. I'll talk more about our family later. Keep reading.

Another story that sticks in my mind happened during one of my niece's high school graduation ceremonies. The class valedictorian, who was Black, attended a predominantly white school and shared this experience. She said, "I remember when I was in sixth grade, my white teacher made me take a test over because I made a perfect score. She stated that I would have to take the test over because she believed I cheated. She told me that 'You people (Blacks) are not smart enough to make a perfect score on any test.' I scored a perfect score on the second test, and the teacher never said a word. I thought to myself, I will show that teacher that she was totally wrong. Look at me now. I am the valedictorian of my graduating class." The young lady received a standing ovation. I loved it. I mean, I absolutely loved it.

I learned throughout my life experiences, as I mentioned earlier, that we should never assume anything about anybody because of their race, the color of their skin, their religious preference, how much money they make, or any prejudices you may possess. If you want facts about a person or a group of people, do like the young lady did, who I talked with in Jonesboro. Ask them. Otherwise, keep your mouth closed.

Anyway, back to recruiters and stereotypes.

The recruiters would schedule the applicant to take the written test and the physical. No completed application was needed for this process. If the applicant passed both tests, we were left with an applicant who was qualified to enlist but couldn't enlist because the recruiter didn't complete an enlistment packet for the application.

What this does is create a statistic called "QNE" (Qualified but Not Enlisted). This was a statistic I refused to accept. I'll talk more about that later. Additionally, it means that there is a person who is interested in joining the Army but for some reason didn't. It also means that when you don't move an individual when he or she is ready, they just may join another branch of service, our competition. Or even worse, the applicants could miss out on training that could have taken them down a profitable career path.

What did I do to fix this problem? Well, the fact that I was not in the recruiter's chain of command did not stop me from correcting a situation that obviously had been allowed to get out of hand. I am really surprised that the company leadership allowed this to continue. I have more about that later. Keep reading.

Each time I had an applicant for prequalification, I talked with that applicant with one purpose. I simply wanted to find out if they were qualified to enlist without a waiver and if they were ready to enlist. My last question to them was, "If I can get your recruiter to complete your application, will you sign a contract today?"

Once the applicant's qualification was confirmed, I would call the recruiter. That's right, I would call the recruiter of record. I guess you ask why not the First Sergeant. Well, first, I never met the company leadership, and time was of the essence. Because the swearing-in ceremonies were scheduled, I didn't have time to deal with the chain of command. I went directly to the applicant's recruiter, and I was very direct.

The conversation went something like this: "This is your Senior Counselor at the MEPS. You have an applicant on the floor without a completed enlistment packet, and they are ready to enlist." I'd continue, "You have one hour to get me a completed enlistment packet, and I will put your applicant in the Army today." Although the MEPS staff oversaw applicant swearing-in, I had no problem changing the number of applicants I had to swear into the U.S. Army.

You see, one of the first things I did when I started this position was establish a good rapport with the MEPS staff. For example, each morning, I would report the number of applicants to be processed for enlistment to the Battalion Operations staff. At the end of the day, the number that I enlisted, most of the time, would be more. I did this consistently at least two or three times per week. I was very successful doing this, and the good thing was that the Battalion Commander and the Battalion Sergeant Major never had a problem with what I was doing. They just reaped the benefits.

The only thing that I was told was that the Battalion Commander asked the Battalion Sergeant Major, "What the hell is Sergeant Wilson doing down there?" The Battalion Sergeant Major replied, "I don't know, but whatever it is, let him keep doing it." And I did. The thing I liked most about recruiting was that if you were successful, you called your own shots. The people upstairs took care of you. Keep reading. I'll talk more about that later.

As time went on, I kept doing what I was doing. I kept putting people in the Army. If an applicant came into my office, they signed a contract. I can honestly say that I never used any type of trickery, and I never lied to any of my applicants or their guardians to get them to sign a contract. I knew my product, and I knew that if an individual made an application, they had an interest in joining the U.S. Army.

Therefore, I capitalized on their interest. I had many applicants who joined the U.S. Army through the Delayed Entry Program (DEP). Some of them then had doubts during the delay period. Although I never expected anyone to back out, I knew it was a possibility. No problem. I always sympathized or empathized with the applicant, depending on their reason. Using proper sales techniques, I always convinced the applicant to complete their commitment. I never lost an applicant from the DEP. I'll talk more about that later.

CHAPTER 27

My Assignment as First Sergeant of the Amarillo Recruiting Company

As I mentioned earlier, my selection for promotion to Master Sergeant (E8) was the reason I was assigned as Senior Counselor in the Amarillo, TX MEPS. I held the position for approximately one year. Sometime around March or April of the next year, I received my promotion, which simply meant that once I was promoted to Master Sergeant, I started getting paid at the higher grade. Man, was that great! (Smile.) The Battalion Commander and Battalion Sergeant Major came to Amarillo to officially promote me.

However, there was a surprise I never saw coming. We had a great promotion ceremony. After the ceremony, we all stood around just sharing recruiting war stories. That's when the Battalion Commander hit me with the shocking news.

"Master Sergeant," he said. "My boss got a call from a higher Command HQ. We could lose you."

At this point, I'm just listening. I learned that when the Colonel is talking, you don't interrupt him.

He continues. "The boss said that if we don't assign you to a position that will allow you to exercise your true abilities, they were going to place you themselves."

I'm still listening. "We have decided to make you the First Sergeant of the Amarillo Company."

He paused. At that point, I stated, "No problem, sir."

The Battalion Sergeant Major jumped in with a "let's see what he says about this" look on his face.

"Now, Hendry, he said, you won't have a Company Commander for a while, and we really don't know when you will get one."

I responded with a totally positive attitude, "No problem, Sergeant Major. I can handle it."

"Ok, Hendry, you start right away," the Battalion Sergeant Major said.

Man, was I shocked. Not having a Company Commander for a while turned out to be a great thing for the company and me. Let me share this quick story before I get back to my First Sergeant/Acting Company Commander experience.

Remember my friend from Jonesboro, Gary Dellinger? Well, he called me from his station in Dyersburg, TN. Although Gary and I spoke on the phone regularly, it was always great hearing from him.

"What's going on?" I asked after a brief conversation.

"Man", he said, "I think I want to be a Drill Sergeant."

"Great," I said.

You see, I talked about my Drill Sergeant Assignment so much that I was not surprised that he chose to try his hand at Drill Sergeant Duty. Plus, it would be another feather in his hat at promotion time. He was accepted into Drill Sergeant Duty and worked there for a couple of years. Gary called me again to let me know that he had been promoted to E8 and wanted to come back into recruiting duty. After congrats, I gave Gary the number to my Battalion Sergeant Major and suggested that he contact the Sergeant Major right away. You see, my Sergeant Major already knew about Gary because I talked about him so much whenever the Sergeant Major and I talked. I was confident that the Battalion Sergeant Major would snatch Gary up, and he did.

Gary became the First Sergeant of one of the two Oklahoma City Recruiting Companies, largely, I believe, because he and I had worked together before, and we were extremely successful. At this point, I was doing great things in the Amarillo Company. I believed that the Sergeant Major was confident that Gary would do the same in Oklahoma City.

It was no surprise that Gary did just that. It so happened that my friend and I were Acting Company Commanders of our respective companies during the same period. Both of us attended a Brigade Commander Meeting as acting Commander. This is a meeting where each Company Commander within the Brigade would brief the Brigade Commander on their Company's successes and failures.

What really stood out among all of the briefings was that my company stats and Gary's company's stats were almost identical. Except for a few numbers here and there, our charts were like a carbon copy. We had not worked together for a long time, but we believed in recruiting and were committed to our customers. What we did as recruiters/recruiting leaders made us extremely successful.

I must say that we were both super great recruiters, super great Station Commanders, outstanding First Sergeants, and great acting Company Commanders. Gary and I, and each of our spouses, Renee and Sarah, respectively, have visited several times by phone and in person since our retirement. Now, back to my First Sergeant/Acting Company Commander experience.

Although it was a shock to me that the Amarillo Recruiting Company didn't have a Commander or a First Sergeant, it was an honor that the battalion leadership had enough trust in me to assign me as the Company First Sergeant and Acting Commander. Although this was my first First Sergeant assignment, not to mention acting Commander, I didn't have any doubt about what had to be done and that I could get the job done. I was an extremely successful recruiter, as well as a Career Counselor and Senior Counselor, and a proven leader with over 15 years of leadership experience while in the U.S. Army, so I just went to work.

As I sat and analyzed my staff and area of responsibility, I couldn't help but notice where the Company was as it relates to our enlistment percentages. The company was in the last week of the recruiting month. Let me explain. Each Recruiting Company receives the following month's mission quota for enlisting applicants into the U.S. Army. The Company Commander and First Sergeant usually assign each recruiter their numbers for the month, which culminates in the station's mission. Having said that, I noticed that my Downtown Amarillo Station had made their numbers but had not enlisted a female, which was in their mission box.

I took the opportunity to go over and meet the Station Commander of my largest Multi-Man Station in the Amarillo Recruiting Company. We sat in his office and just chatted for a few moments before I got to the real reason for my visit. Initially, I approached the subject in a manner to give the Station Commander the opportunity to make the decision himself to process the applicant. Since he was insistent on not processing the applicant, I simply directed him to process the female applicant right away.

The conversation went: "Sergeant, apparently, you don't understand what I'm saying. I am not asking you to process this applicant; I am ordering you to put that female applicant on the floor for enlistment tomorrow."

The Station Commander stood up and slowly walked out of his office, into the station's open area, obviously pissed, and stated in a loud and unprofessional tone, "This is my station, and I will do what I want with my applicants."

At that point, I became extremely angry instantly. People have told me that when I am angry, I have an expression that will make a gorilla piss. As I approached the Sergeant and stood directly in front of him, I said, "What did you say, Sergeant First Class?"

The Master Sergeant from Battalion Headquarters, who was conducting a station inventory, quickly jumped in and said to me. "Hey, Top, let's go get a cup of coffee."

I don't know if we ever got a cup of coffee or not. I do know that I went to my office right away, wrote the Station Commander up, and called him into the office. I had the Sergeant sit across from my desk as I explained a few things to him.

I remember starting with, "Sergeant, one thing I do not and will never tolerate is insubordination. If you refuse to do what I direct you to do, I don't need you as a supervisor in this company. If you insist on refusing to do what I tell you to do, I want you to report to Battalion Headquarters on Monday morning."

The soldier chose to leave my office. I quickly looked at that station's recruiters to see who I would place in charge of the station. When I called to speak with the senior sergeant, the previous Station Commander answered the phone. I asked, "Why are you in that office, Sergeant? Pack up your stuff and leave the office. I don't want you there."

I just didn't want him to influence the other soldiers. I appointed the senior recruiter and briefed him on what I wanted him to do.

On the following Monday morning, the Battalion Commander and Battalion Sergeant Major called and asked me what had happened. I explained that the Station Commander disrespected me and was insubordinate. He refused to do what I ordered him to do. The Battalion Commander stated, "Did you have to send him to the Battalion?"

I responded, "Yes, sir. I have a Recruiting Company that's in trouble that I have to get on track quickly, and I don't need the second senior soldier in my company undermining me. I didn't want him around my soldiers. I thought the best thing was to send him to the Battalion for reassignment."

The Battalion Command upheld my decision, and I really appreciated that. They sent me a new SFC Station Commander in a month or so. In the meantime, I moved forward with what I had to do.

During the first week of my assignment, the Battalion Sergeant Major accompanied me on my first visit to each recruiting station, which was extremely helpful to me. He was very clear about the Battalion Commander's expectations of the Amarillo Recruiting Company. I thought to myself, man, this is easy. Those are my expectations as well. My company's area covered the entire Texas Panhandle, the Oklahoma Panhandle, and several counties in the southwestern-most corner of Kansas. Needless to say, I clocked several hundred miles during that assignment.

I continued to study my area of responsibility for my company, which included a breakdown of each recruiting station's area of responsibility. This was extremely important for the success of my company. At the same time, I wanted to know the grade and how many recruiters I had in each station.

That's all of the information I wanted to get from each station's administration files. Immediately, I conducted my first company meeting, at which time I gave a synopsis of myself and what I expected from each of them. Within a week or so after my assignment, I conducted an on-site visit with each recruiter. I made mental and written notes of my findings— good and not so good—which I addressed during my first company meeting.

Shortly after I visited each station, I conducted another very important company meeting. This meeting was critical because there were a few points I wanted to instill in each of my recruiters. First, honorably enlist qualified applicants in the United States Army. My point was that I would not tolerate fraudulent activity. As I continued, I told them to be respectful of each applicant and his or her family during their recruiting efforts and to never, ever give false information to a prospective soldier or their family.

I also wanted to instill in each member of the Amarillo Recruiting Company that it is important to know what the Army offers each recruit once they become a soldier. I said to them, "I know that most of you have only served in one career field before recruiting. Therefore, it is imperative that you learn as much as you can about other career fields within the United States Army and what jobs each career field offers."

I went on to stress the importance of never misrepresenting our Army with false information to get a contract. I assured them that I would always be available if they needed help closing the sale with an applicant. I tried to instill in each of them that if the two of us can't convince the applicant to join the U.S. Army, we will ask them for a referral. Also, I insisted that they always follow up with that prospect in a timely manner.

Finally, during my recruiting experience, I confirmed that once some recruiters made their current mission for the month, they would hold applicants for the next month's mission. I explained to my team that in no way would I allow sandbagging to continue during my watch. Each of your applicants deserves the right to sign up to serve the U.S. Army when they want to, not when you want them to. I continued to reinforce my training during this assignment. Fortunately, I had a group of great young professionals. They were disciplined soldiers who understood my demands and acted accordingly.

I visited each of my recruiting stations regularly. Sometimes we would just sit and talk about anything but recruiting. However, I spent a lot of time training new recruiters and reinforcing my demands to the older guys. I had no problem with my staff, which was proven through our production. I'll explain later. Keep reading.

One concern that kept coming up during the meeting was that the Battalion Command would not allow a recruiter to enlist applicants who were not in their box. Let me explain. Each month, each recruiter is assigned a mission consisting of a number of citizens. These citizens are assigned by type—male, female, and by mental category 1, 2, 3, or 4. I told my guys, "You just put qualified people on the floor. Let me deal with the Battalion. That's my job."

As I continued to explain, "If you have good prospecting ethics, you will, without a doubt, contact more people who are qualified in categories 3 and 4 to join the Army than those who are in categories 1 and 2. The Recruiting Command will allow all qualified applicants to join the Army at some point during the month. I will do my best to get the Battalion to allow all qualified applicants we put on the floor to enlist. Trust me, it will all come out in the end.

"If you keep prospecting, by the end of the month, or quarter, or year, not only will you make your mission, but there is a good chance that you will also make mission by category. Let me leave this recommendation with you. Once you make your numbers, concentrate a little more on the types of applicants that are a part of your mission. However, never neglect any qualified applicant. Always process any qualified applicant, regardless of their gender or test score, right away. Put them in the Army on the very first day possible."

After a really long meeting, we were off and running. Man, did we put some people in the Army my first year as First Sergeant. Remember, I was assigned to this position with no Company Commander in sight, which was a blessing for me. Since I was the Acting Company Commander, this gave me the opportunity to do what was necessary to get the company on the right track. Once again, I did my thing. Not only did the Battalion Commander and Battalion Sergeant Major take notice, but the First Sergeants of the other four companies took notice as well.

I found out that recruiting was very competitive within the battalion. Fortunately, I didn't pay any attention to that, and I didn't want my recruiters to get hung up on counting percentages. My motto was, "Just continue to put qualified people in the U.S. Army and see what happens in the end."

Before my Company Commander arrived, I had the opportunity to be interviewed by a local TV station. I was also contacted to represent the U.S. Army during a Memorial Day celebration. The local news simply wanted to do a segment on the number of local citizens volunteering to support our country's efforts during the time of war. The interview went well.

My part during the Memorial Day celebration didn't go well. I arrived at the location prior to the time I thought I was given, and guess what, the ceremony was over. I stood there dressed in my Class A uniform, looking for my contact person. When I saw him staring at me from a distance, I did a "what's up" gesture, and he turned and walked away, so I did the same.

I don't know if he gave me the wrong time or if I recorded the wrong time. I really hated that, but I took responsibility. To this day, I check times and dates and verify that they are correct for everything I am involved in. I also always arrive fifteen minutes or more prior to the time I am supposed to be there. That's exactly what I thought I was doing for the Memorial Day celebration. But I failed.

CHAPTER 28

The Company Commander Arrives

By the time my Company Commander was assigned, I had things going my way. All he had to do was take care of the Company Commander's stuff. This included knowing the statistics by station and recruiter, which gives you the company statistics. Then he would report the same to the Brigade Commander. Let me tell you something, this man was a Godsend.

I introduced him to the recruiters as we visited each of the recruiting stations throughout the company geography. I informed him that we had a bunch of great Army recruiters. "Sir, I will continue to train them if you will take care of their personal, professional, and family needs. As a company, I think we will make mission plus more each month."

My Commander stated, "I can do that, Top."

I really liked him from the start. He was an E6 Noncommissioned Officer in the U.S. Navy who had been accepted into Officer Candidate School, which I think was a plus for the U.S. Army. I didn't know if this was his first assignment as a Company Commander, and I really didn't care. After I briefed him while sitting in our office, we toured our recruiting stations.

After meeting each recruiter and receiving a Station Commander's briefing, he couldn't stop smiling. If I could read his mind, it seemed like his thoughts were, *I think I fell into a gold mine.* After completion of the station briefing, he commented that all of our recruiters were doing well in their recruiting efforts and that they seemed well-motivated. During the next year, yes, year–our Company Commander held up his end of our plan and just reaped the benefits. Man, did we gel as the Company Leadership Team.

After a year, our Company Commander was reassigned to the Battalion Headquarters. I will have more about that later. The Commander and I worked extremely well together, as did our wives. The Commander took care of his responsibilities, and I continued to do my thing. His arrival gave me the opportunity to concentrate solely on my men and their sales techniques. I also got the opportunity to spend more time with my lovely family.

Our two daughters, Dimey and Dana, were doing great in school. They were very active Girl Scouts in our community as well. In the meantime, my wife and I decided to try for a third child, whom we hoped would be a boy. With God's blessing, on June 21, 1983, Sarah gave birth to a lovely little boy, Henry Clarence Wilson Jr. (we call him Clarence). He is now married to a lovely lady, Shannon Arnold Wilson.

Clarence and Shannon were blessed with three beautiful children: Henry Clarence Wilson III, Patrick James Wilson, and a recently adopted beautiful baby girl whom they named Iris Marie Rose Wilson. Henry III is seven years old and is in first grade, where he attends Holy Trinity Montessori School. Clarence works for the US Federal Government as a manager, and Shannon is a very successful CPA for a large medical insurance company, EmblemHealth, where she is the "Financial Analysis Leader." Both are doing extremely well in their respective jobs.

My wife and I are glowing with joy as we can babysit Henry Clarence Wilson III after school. Yes, we also have the pleasure of spoiling Iris and Patrick each weekday as well (Smile.). The parents decided to call him—wait, wait, wait, let me back up a bit. When Henry Jr. and his wife were discussing naming their new baby Henry, they had several questions.

My son asked, "Dad, is there a rule that the first son should be named after his dad?"

I stated, "As far as I know, there is no such rule. I have noticed, not only within the Black race but in other races as well, that there are parents who name their firstborn son after the father. I'm sure not all firstborn sons are named after their dad. You do what you and your wife want."

Later, maybe a day or so later, my son called again. "Dad, would you have a problem if we name our son Henry after me?"

From the excitement, I felt a warmth rush through my body. I smiled as I answered, "Son, it would be an honor to have Henry Clarence Wilson III in our family."

You may have guessed by now that we have Henry Clarence Wilson III.

"So, what do we call him?" my son asked.

"No problem," I said. "Some people call me Dad, some call me Uncle Henry. I was called Hank when I played baseball. My wife calls me Henry, honey, or some other names I don't want to list. (Smile.) My family used to call me Henry Clarence back in the day. Since we call you Clarence, why don't we call your son Baby Henry?"

His mother and father agreed. To this day, he responds to both Henry and Baby Henry. I like that because I get to sandbag when my wife calls out for Henry. I never answer. (Smile.) Yes, you guessed it, she gets what she wants in the end.

My wife, now a mother of three, continued to do well in her government job. During my First Sergeant assignment, she was accepted into the management program with the Treasury Department. Sarah received acting assignments mostly around the Dallas area when we lived in Amarillo, TX. Fortunately, we found a very reliable babysitter, Michelle, who was recommended to us by some of our very good friends, Leotis and Sheila Dixon, Michelle's sister and brother-in-law. We met the couple earlier during my assignment as Senior Counselor.

Michelle was so caring and responsible that we never had to worry about our children; well, no more than any caring parents would. Michelle would take our girls to school and pick them up when needed. I continued to travel throughout my area while Sarah was on acting assignments. Most of the time, I would visit my out-of-town stations on a one-day turnaround trip, which would get me back home during the early evening. Michelle would take care of what was needed while I was away. I would always schedule overnight visits to my Kansas Recruiting Station when Michelle was available to stay with the kids. I'll have more about the Dixons later.

CHAPTER 29

One Year as First Sergeant

My first full year as Acting Commander and First Sergeant of the Amarillo Company, my company smoked the other recruiting companies within our Battalion. It seemed that my recruiters were just waiting for my kind of leadership. They knew that I was a serious disciplinarian and that I was very knowledgeable about the recruiting mission. They also knew that I cared about each of them and would do what was necessary to take care of them. I stressed to each recruiter to never neglect their family.

My Company Commander never got in the way of what I did to get the job done. He always backed my decisions, and I supported my Company Commander to the fullest. Working together as a team made it extremely easy to manage our company. Our recruiters seemed very happy doing what it took to put qualified applicants in the U.S. Army.

Going into my second year as First Sergeant of the Amarillo Recruiting Company, my Company Commander was reassigned as Chief of the Battalion Operations Section. In his new position, he was responsible for the Battalion's recruiting statistics and training of all Battalion personnel. Of course, he was available to support the Battalion Commander whenever he was needed. I also believe that the Battalion Leadership was hoping that he could incorporate some of the things we were doing in our company to be successful into the Battalion Training Program.

The most important thing that my Commander's reassignment did for the Amarillo Company happened toward the end of my second year. Let me explain. I told you earlier that my company advanced from last to second place according to the Battalion Recruiting Statistics during my first full year as First Sergeant. When my Company Commander was reassigned as Chief of Battalion Operations, he placed a lot of emphasis on the statistics for each of the Battalion's recruiting companies.

In my position, I only concentrated on my company's numbers to ensure we were on target to make mission each month. Well, each company's statistics were a big part of the Battalion Operations Mission. They were tasked with reporting the numbers to the Battalion Commander and Battalion Sergeant Major at the start of the workday and at the appropriate time during the day. This continued until the close of business each workday. The last day for all statistics to count toward the mission and awards was by midnight on the last recruiting day of each month and the current fiscal year.

This is how my previous Company Commander's assignment as Chief of Battalion Operations was a blessing for the Amarillo Company. I explained all of that to say this: Check this out. During the early evening on the last day of the fiscal year, my previous Company Commander called me, sounding extremely excited. He told me, "Top, I found out how the Lawton Company won the Company of the Year last year."

"Pray tell, sir," I said.

The captain continued, "Last year, the Lawton Company was trailing Amarillo in points during the last couple of hours of competition. They put enough paper soldiers on the floor during those last few hours of the fiscal year to take the Top Company Award. We actually had them beat by quite a few points before they pulled that stunt," he said.

He continued, "They tried to do the same thing again this year, but guess what; I cut them off at the pass. Once again, they sent in enough paper soldiers to take the award again this year. However, I told them, "No, you are not going to underhandedly take the award again this year."

I know you ask, "What is a paper soldier?" Well, a paper soldier is a veteran who still has a remaining obligation left on his or her six-year military obligation. The process to enlist them into a reserve unit only takes a fraction of the time that it takes for a non-veteran. An enlistment form is completed and sent to the Army Counselor at the Military Entrance and Processing Station.

Let me make it clear that the captain would not have denied the enlistments if the company needed them to make their mission for the month. However, just like the previous year, they only needed the enlistments to take the Company of the Year Title from the Amarillo Company. I have to tell you that I was really happy to get that news. I thanked the captain for looking out for the Amarillo Recruiting Company and moved forward with my work.

CHAPTER 30

Amarillo, the Company
of the Year

Sometime during the next day, the Battalion Commander and Battalion Sergeant Major called to officially inform me that Amarillo was the Honor Company of the Year. I immediately informed my recruiters and my wife, Sarah, who was on an acting assignment. I have got to tell you, I couldn't wait for the Annual Conference. My soldiers would finally get the notoriety that had eluded them for a long time.

Each year during the Battalion Conference, all assigned soldiers attended a very important meeting conducted by the Battalion Commander, Battalion Sergeant Major, and other key officers and noncommissioned officers from the battalion staff. The instructors would highlight what we, as a Battalion, did right and the areas we needed to improve in. The highlight of the conference was the Battalion Banquet. During the banquet, everyone had dinner, we danced, and most of all, my company, the Amarillo Recruiting Company, was showcased as the Battalion Company of the Year.

When the Battalion Commander made the announcement, "Amarillo Company members, come forward," I motioned for my recruiters to follow me as I took a position in front of the dais. I noticed the proud expressions on my men's faces as the Battalion Commander and Battalion Sergeant Major presented the Honor Company Trophy to me. I was proud of my recruiters as all of us stood there in front of our spouses, all the other recruiting company recruiters and their spouses, and all of the Battalion Staff members and their spouses. That made me extremely happy.

Wait a minute, let me back up a little. My wife flew into Oklahoma City for the conference from her acting assignment. What was great, other than seeing my beautiful wife after a week away, was that I was given a chauffeur-driven limousine to pick her up from the airport. We also received a lovely suite for the conference period. The conference was extremely special for my wife and me, as well as for all the Amarillo Recruiting Company staff and their families.

CHAPTER 31

Sarah's Selection as Tax Auditor Manager

Shortly after the conference, Sarah was notified in December 1986 that she was selected for a Tax Auditor Manager position in Knoxville, TN. Unfortunately, her reporting date was in the middle of the school year—January 1987. Therefore, we had some decisions to make. For me, it was a no-brainer. Without hesitation, I advised my wife to accept the promotion.

However, Sarah had a difficult time leaving our children. We had already decided that we would never take our children out of school after a substantial portion of the school year had passed. Sarah is such a caring mother, and the fact that we had never been separated for that length of time made it hard for me to convince her to leave. Finally, she decided to go only after I convinced her that I would take excellent care of our children. She took the job, and I promised to drive her to Knoxville in early January 1987. I really couldn't believe what happened next.

About two days before we were to leave for Knoxville, I received a call from the Battalion Operations Officer. "First Sergeant Wilson, you have a Company Commander's Meeting in Kansas City, KS, in two days."

Would you believe I had to leave the same day my wife and I were supposed to be leaving for Knoxville? I didn't know if I was more disappointed than pissed off or more pissed than disappointed. I felt bad for my wife because she was obviously disappointed. Sarah had never driven even a third of the distance she had to drive to Knoxville from Amarillo alone. Plus, it had snowed a lot a few days earlier, and snow was still on the interstate.

We both left the same day. Sarah left driving to Knoxville, and I flew out to Kansas City. I flew into Dallas, where it was still snowing and cold. I don't like flying anyway, so when they had to de-ice the planes and then had to pull off the runway and de-ice the plane again, I was so scared that I almost shit on myself. To make matters worse, as we ascended into the air, the airplane shook like the wind was going to blow us right out of the sky. After we leveled off, the rest of the trip was okay. I just didn't want to be up there.

Then, as we were landing in Kansas City, I didn't see anything but snow on the ground. I just closed my eyes and prayed that the landing would be safe. Of course, we landed safely, and our meeting started around 1 p.m. that same day.

We didn't have cell phones then, so I had not talked with my wife since we left home that morning. Suddenly, the phone in the meeting room rang. I was surprised when the Operations Officer said, "First Sergeant Wilson, telephone."

My heart almost jumped out of my chest. As I answered the phone, I could hear my wife on the other end of the phone sobbing. That was probably the most painful feeling I have ever had. I felt so helpless.

After we talked for a few minutes, Sarah calmed down enough for us to decide what to do. You see, she had driven over 280 miles from Amarillo to Oklahoma City in the snow, using the tracks of 18-wheelers, which was a smart move. I suggested that she call Gary and Renee Dellinger and stay over in Oklahoma City before going on to Memphis. Sarah responded, "No, I need to be alone so I can think."

I just said, "Okay, honey," and gave her the name of the hotel I always used. "Call me later tonight. I love you."

She said, "Okay," and we hung up.

I found out after we talked later that evening that my baby was still struggling with leaving the children. My Battalion Commander, who had just been assigned, asked me, "Is everything okay, First Sergeant?"

I responded, "Yes, sir," and the meeting continued.

CHAPTER 32

Changes within the Battalion Leadership

The Battalion Commander was promoted and became the Deputy Commander of the Southwest Recruiting Region. The Battalion Sergeant Major retired and, from what I understand, hooked up with the First Sergeant of the Lawton Recruiting Company. Remember, that's the company that stole the award from me, my first year as First Sergeant. I really enjoyed working with those two professionals. Kudos to them.

Quick Story

Once when the Sergeant Major visited me, we traveled to the Liberal, Kansas, Station. During our conversation as we traveled back to Amarillo, he asked me, "Hendry, do you talk much with the First Sergeant of the Lawton Recruiting Company?"

I wondered where he was going with that question. "Yes, every once in a while, Sergeant Major."

"What do you think of her?" he asked.

Fortunately, I gave an honest answer, which turned out to be the correct answer. Through gossip, I found out sometime later that he and the First Sergeant were courting. I am so glad that my answer to the Sergeant Major's question was on point.

I got along well with all of the First Sergeants. The Lawton, Oklahoma, Company First Sergeant really didn't do anything wrong when she stole the Top Company Award from me. She was just a strong competitor who did what she had to do to get that trophy.

Our friends, the Dixons, moved to Oak Ridge, TN, and my lovely wife was doing great in her new position in Knoxville, which is about 30 miles from Oak Ridge. Our three children were doing well in school and at home. Since my soldiers continued to do extremely well enlisting people into the U.S. Army, and the fact that my new Company Commander was assigned, gave me the opportunity to spend less time at work. So, I spent every minute I could with my children. I also made sure that our children spoke with their mother every day.

In June of 1986, we moved the children to Knoxville with Mom. During the next six months, I requested a transfer to be close to my family. In the meantime, my wife and I sold the house in Amarillo and purchased a house in Knoxville so that they would be more comfortable until our family was together again.

In the meantime, I was selected to attend the United States Army Sergeants Major Academy. This is a six-month U.S. Army Senior Manager Course for senior enlisted soldiers. I was excited and scared at the same time. Successful E8s and E9s are selected to attend the Sergeants Major Academy. I'll have more about that later. Keep reading.

Sometime during September or October 1986, I received a call from a member of the Brigade Commander's Staff asking if I wanted to take the Senior Counselor's position in the Memphis MEPS. I knew that a six-hour drive between Memphis and Knoxville was a lot closer than a 20-hour drive from Amarillo, but I wanted to be as close as possible to my family.

Although I really wanted the Senior Counselor position in Memphis, my family was my motivation. Therefore, I opted for the First Sergeant position in the Nashville Recruiting Company with a reporting date in early January 1987.

I informed my staff of my departure plans, and I can honestly say that the atmosphere was very somber. Without a direct question, I read their expressions to say, "Damn, First Sergeant, why do you have to leave now?" I believe it was because I had brought them from last place in the Battalion to first place, realistically, in two consecutive years. All I could do at that point was to instill in them to continue to do what we had been doing for the last two and a half years, which was to continue to work hard and to continue to be honest with their applicants and their families. I believed that would, without a doubt, secure each recruiter's future in recruiting and in the U.S. Army.

My staff was very generous with my going-away party. I had a great time. At the end of the day, I let each of those guys know how much I appreciated their support during my tour as First Sergeant and Acting Company Commander. They were simply great. I had no problems with them responding to my tough demands. I have no doubt they appreciated the training I instilled within each of them, the direction I gave, and my relentless support that was never denied.

I also received a great going-away gathering with my counterpart First Sergeants from the other four recruiting companies. It was just what I needed since each of them—except my friend Gary from Jonesboro—was so competitive. Don't get me wrong, competition is a great thing if it's done above board. My friend Gary Dellinger and I had similar attitudes. It's simple: just do your job.

I really enjoyed the gathering. It was great to share a kind of emotional feeling of camaraderie with them. We just sat around, eating, drinking a little spirit, and sharing recruiting war stories. What really surprised me was when one First Sergeant stated, "Well, Henry, I really don't mean any disrespect, but I'm glad you're leaving."

"Are you kidding, man? Why is that?" I asked.

"Well," he continued, "with you gone, maybe I can win Top Recruiting Company occasionally." I could see he was very serious because he never cracked a smile. All I could do was laugh and thank him for the compliment.

CHAPTER 33

My Departure from my Amarillo Assignment

As I traveled across the country on I-40E toward my new assignment, I thought about how I would make another failing Recruiting Company successful. I felt extremely confident about the assignment. As I continued, very pleasant thoughts crossed my mind. For the first time, I realized that I only had six months left before I could retire with twenty years of active service.

I thought, *Man, that's great.* But that thought went from my head as fast as it came in. You see, I never dwelled on the possibility of retirement. Anyway, the fact that I was selected to attend the Sergeants Major Academy locked me in for several more years because I expected to be promoted to Sergeant Major and, of course, take a position, hopefully, in the recruiting command.

First Sergeant: Nashville Recruiting Company

The First Sergeant assignment in Nashville brought me to a low that I had never witnessed before. This is true not only during my Nashville period of service in the United States Army, but during my prior and post-Army life as well. As I said earlier, I grew up poor. I don't recall any days when I went to school with money to buy lunch, except, of course, between the ages of 16 and 18.

I remember growing up in a segregated society where, at one time, I was picked up and taken to jail with no explanation, simply because I was Black. I remember having to hide in the bushes at night as I walked home because people would throw objects at me from their cars as they passed. I remember being taken advantage of sexually by schoolteachers more than once. Even after my Army retirement, I remember being discriminated against by at least one Black manager. I guess you are wondering where I am going with this. Well, with all these experiences, after a while, I understood what was going on.

The fact was, being poor and growing up in a segregated society is a no-brainer. That's life as it were. Being sexually assaulted in high school, especially by teachers, was something I didn't expect, but I understood what was going on after a while, and I was able to fix those problems. I was really surprised to be discriminated against by a Black manager more than once during my employment with the federal government.

No problem, I fixed those problems by continuing to do my job, plus any additional work I was asked to complete. In addition, I did anything and everything I could to bring attention to a very ineffective manager. However, what went on in my Nashville First Sergeant assignment, to this day, has been an unforgettable mystery to me.

After the third month of the assignment, I knew there was something strange going on. However, unlike any of my assignments during my past 19-plus years of work in the U.S. Army, for the life of me, I could not figure out what it was. Let me explain.

I was so excited to be assigned to the Nashville Recruiting Company. Of course, I did research on the company's statistics for several previous years and found out that the company had been failing for a while. I was told that the First Sergeant, prior to my assignment, also Black, was relieved. No problem, I thought. After all, I had always been given the tough assignments. This was right up my alley.

My immediate supervisor, the Company Commander, prepared a reception for me at his apartment. I had the opportunity to meet all of the company recruiters and the Company Commander's spouse. The food and the presentation were excellent. The Company Commander took me to all of my recruiting stations to meet the recruiters. I was surprised, however, that the Brigade Sergeant Major did not take me around or didn't accompany us. That became a red flag to me later. Keep reading.

The Company Commander gave me a synopsis of each recruiter as we were en route to their station. This is something that always happens when there is a change in leadership. However, I never let it affect my mode of operation. I judge each recruiter on what I see as it relates to his or her performance and appearance.

I know that leaders have different leadership styles; therefore, subordinates react to what's being projected by their manager. A manager cannot and should not manage a member of his or her team based on the recommendation of the person he or she is replacing. As a manager taking over any element of an organization, you owe it to your people to be fair and impartial.

As we traveled to each station, two things stuck out in my mind. I looked at the statistics in each station. The stats on one station stood out to me. I noticed that the recruiter was heavy in high school seniors. Recruiting seniors is admirable, but if you don't manage the program properly, neither you nor the Army will benefit from the program. The recruiter's statistics also showed me that he was not effectively covering the rest of his recruiting population.

I stated to the recruiter, "I'll be riding with you soon. I'll show you how to recruit."

He replied with an attitude, "I know how to recruit, First Sergeant."

I went back to the statistics map just to send the recruiter a message. I looked at the map briefly, turned and looked back at him, smiled, and the Company Commander and I left. I never got a chance to physically visit that recruiter, as I intended to do with all Station Commanders. What I noticed while viewing the map was that the recruiter was working his high school market very well.

Don't get me wrong, this is a great thing. Recruiters must really stay in contact with their seniors during their senior year of high school to keep their minds focused on graduating and leaving for active duty on their scheduled departure date. What's even more important is that the recruiter should instill in the applicant that signing the Delayed Entry Program contract obligates them to spend a number of years in the United States Army. However, what I suspected at first glance of the station's recruiting map came to fruition.

Sometime toward the end of the school year, I started receiving calls from the recruiter's high school seniors who had joined the Delayed Entry Program. Each student would simply give me his name and proceed to tell me that he or she had changed their mind about going into the Army. Of course, I tried to convince them to follow through with their contract.

However, we had no choice but to release them from the program. To really be successful in recruiting, recruiters must work their entire market, just in case they are faced with those types of situations. The recruiter in question left recruiting shortly after I was assigned. One of the few times I spoke to my Battalion Sergeant Major, I explained this situation to him.

His response was something like, "You are trying to blame the recruiter for your failure, First Sergeant?"

Of course, I didn't expect anything more from him. Based on my short period and experiences with the Sergeant Major, I chose not to respond. Let me back up a little. As I continued to visit my recruiting stations and communicate with my recruiters, I learned that I had a group of professionals who could do the job.

Of course, some were better than others. Some seemed stressed and burnt out, and unsure of themselves. During the first three to four months, I had serious conversations with some of the recruiters who seemed to be struggling with their mission. They seemed to have gotten my message and showed signs of improvement. However, the enlistments just were not there. I dug deeper into the Recruiting Station statistics and the recruiter's administration to try to get an idea of what was going on with each recruiter.

For the most part, the recruiters were prospecting and making enough contacts that they should get them enough appointments to make the mission. However, the appointments just weren't there. No recruiter will make mission consistently if they do not make and conduct appointments. Therefore, I instituted a program that I developed during my assignment as First Sergeant in Amarillo called the "PET" Program (Prospecting for Enlistments by Telephone).

What this program did was require each recruiter to work in his or her office from 6:00 p.m. to 8:00 p.m. on Tuesday and Thursday evenings. During that period, each recruiter, using any administrative document at their disposal, was required to call prospects to secure at least two appointments on each of those days. Each Station Commander was required to call me to verify their station's outcome for each of those two days.

Once I saw the necessary appointments where the company should make mission, I would stop the program and focus on conducting appointments. I used this program several times during my last assignment, and it worked perfectly. However, in Nashville, for some reason I could not explain, things did not seem to be improving. The appointments were there, but the enlistments were not. It was to the point where, for the first time during my military career, I began to feel the stress of this assignment.

During a few months of this assignment, things that I could not explain really began to get to me. Each evening, I would go to my apartment, sometimes after approximately 7:00 p.m. Usually, I would sit and have a drink to relax after a long day. However, things got so bad that I would not drink when I got to the apartment. I wanted to keep my head clear as I tried to figure out what in the hell was wrong with the company. Why weren't we making mission? I asked myself, "What am I doing wrong?

I shifted gears and started working with Station Commanders who had the rank but were not producing. One, who was the company's top recruiter at one time, so I was told. I noticed that he had not enlisted anybody in the Army since my assignment. I rode with the recruiter, covering a large portion of his area. We visited post offices and Centers of Influence locations. The lack of updated advertisements (pamphlets) in the areas we visited indicated that this recruiter had not visited his area in quite some time.

Each recruiter has Army pamphlets stamped with his or her contact information. Because of what I found, or didn't find, I did everything I could to maintain my cool as we traveled back to home base. When we returned to his station, I took a good look at parts of the office administration, specifically the Lead Source Analysis Worksheet.

I was kind of shocked, but not really, to learn that everybody who enlisted in the U.S. Army from that station during the month prior to my arrival was a "walk-in." Don't get me wrong, walk-in applicants were always welcome. However, it is absolutely essential that recruiters use all tools that are provided by the Recruiting Command to continue to prospect each day. They must not rely on walk-in traffic alone to make mission.

What really pissed me off with this guy was his lackadaisical attitude when I approached him with what I had found. The training I had conducted with the recruiters, the conversations I had with this Station Commander in the past, together with this visit, I determined that he was not equipped to work alone. Of course, the fact that the company was not producing didn't help with my frustration. I immediately wrote the recruiter a counseling letter with the intent of moving him to someplace where he could be supervised. Here comes another shock.

Within an hour, the Battalion Commander hand-carried a letter to this recruiter countering what I had done. That was another red flag that I took notice of. Based on my experience, Battalion Commanders just don't do stuff like that.

The fact that the Battalion Sergeant Major and my Company Commander never talked to me directly to voice their concerns and to offer assistance told me that something was going on that I couldn't physically see. This caused my frustration to continue to push my stress level higher. My feeling that there was something clandestine going on to cause my company to fail was obvious, but what that was escaped me. Factually speaking, whenever the Battalion Sergeant Major spoke with me, he spent most of the time complaining about two senior soldiers who I didn't know at the time, who were Black.

One of the soldiers he accused, without justification, of transporting drugs to Nashville from Florida. The other soldier was the First Sergeant of the Chattanooga Recruiting Company, who fortunately got promoted and transferred out. The Battalion Sergeant Major referred to each of them as "no-good". He said they were like two crows sitting on the stoops, complaining. To be honest, I didn't want to hear any of that nonsense. For the first time since I have been in a supervisory position, I felt this fear that a successful future in this position was in jeopardy. I believe that this is when I let this fear take control of my mental and physical abilities.

Shortly after this episode, I visited one of my multi-man stations. After I completed the business portion of my visit, we just sat and talked, just as I did most of the time when I visited my Station Commanders. As we talked, the Station Commander stated, "Top," he paused for a moment, "you need to watch your back." That's all he said. That was enough to all but confirm my suspicions. Shortly after my visit, I witnessed this strong but negative feeling in my body. It was like my body was trying to shut down. The feeling was such that I had to find medical attention immediately.

When the nurse took my vitals to the doctor, he immediately came to the waiting area to get me. The first thing he said was, "Mr. Wilson, I'm Dr. G. What have you been doing, man? Your blood pressure is out of control."

I immediately broke down with uncontrollable tears. All the pressure that had been building up within me for at least three months caused me to just explode. He gave me a dose of blood pressure pills and a prescription. We sat and talked for a while, which was good for me because I got a chance to talk about things that I needed to get out.

Shortly after that, I called another training session. I refused to give up even though I had a feeling that my days were numbered. I would admit, if those feelings came to fruition, I actually looked forward to leaving. I really wanted to make that company successful.

However, at that point, and with my perceptions running wild, my health was more important to me than anything else. My efforts were always placed on doing the best job possible, no matter where I was assigned. I knew this assignment was a fluke. Something was going on to make me fail. But I just could not figure out what it was and why.

Anyway, just like every meeting I conducted with my NCOs, the Company Commander was always there. For 20 years of service, that had never happened to me. Understand, however, any Company Commander always has the right to attend any meeting called within his or her command area, but they don't normally attend an NCO meeting unless they are invited.

I called the meeting because I had NCO business that I wanted to talk over with my recruiters. The main thing was that I wanted to know why several recruits were scheduled to be processed that morning, but not one applicant showed up at the processing center. I mean, it was kind of strange that there were, I believe, something like 10 to 15 recruits scheduled for processing, but none showed up.

Of course, I just wanted to talk with my recruiters, but I didn't because I never trusted that Company Commander from the day he was assigned. Let me digress a little. Once, I was going on an appointment with one of my Station Commanders. The Company Commander decided that he would go as well. I wondered why, but no problem, we decided to meet at the office and ride together to the site of the appointment.

When I arrived at the office, he had already left, I assumed for the appointment. So, I went to the appointment site, and there he was sitting with the recruiter and his applicant. Man, was I pissed. I sat for a minute and left. I started to talk with him about that later, but thought better of it. Anyway, back to the meeting.

At the end of the meeting, I asked the Company Commander if he would get me a meeting with the Battalion Commander. I don't know what I was going to say. I guess I just wanted to get some things off my chest. Well, it was just as well that I didn't know what I was going to say because I didn't get a chance to say anything. The Battalion Commander fired me when I walked into his office, and I turned and walked out. I was pissed and mentally relieved at the same time.

I started the rebuttal process right away. Why, I really don't know. I guess I wanted to show that I didn't agree with the relief action levied against me. I wanted the records to reflect everything. As I visited each level of command and listened to what was and what was not said, I gained knowledge that put some things into perspective.

My meeting with the Brigade Commander was kind of unnerving. He had his aide taking notes or pretending to take notes as I spoke. However, he seemed not interested in me or what I was saying. He stood up and almost ended the meeting before I finished speaking. I don't remember what I said, but I said something as he stood up to leave. As he and the aide walked away, he turned and snapped at me, saying that he was not going to change the Battalion Commander's decision.

He acted as if he knew me from the past. I programmed his actions in my mind and proceeded back to Nashville. As I connected at least two red flags, I believed that the Brigade Commander had ordered my relief. I said to myself, "No problem, forget it. Let's see where this takes me." To no avail, I continued the rebuttal process all the way to the Commanding General of all of the Recruiting Command.

As I prepared to leave the command headquarters, I ran into my past Brigade Commander (when I worked in Amarillo), who had just been promoted to Brigadier General. "Hello, sir", I said.

"Hello, First Sergeant." The General responded.

"Congratulations on your promotion, sir", I said.

He replied, "It's soldiers like you who got me this promotion. Thank you."

"It was truly a pleasure working for you, sir," I stated as I proceeded to leave the building.

I found out that my present Brigade Commander was also prompted to Brigadier General. Knowing that, I thought to myself, *You know, Henry, it may be best that you are leaving the Recruiting Command. You know, for whatever reason, my present Brigade Commander, the new General, does not like you, and you could end up getting an assignment on his side of the Recruiting Command.*

As I cleaned out my desk, the Company Commander came into the office, walked past me, then came back to my desk and asked me, "Hey, First Sergeant, you know who is taking over Command of the Military Personnel Record Center?"

I couldn't care less, but I answered, "No, sir, who?"

He continued, "Our Recruiting Command Commander."

With a conniving grin and a sneaky chuckle, he turned and walked away. Any fool could tell that wasn't the real message. What he was really telling me was that, with that General sitting in the Military Personnel Records Commander Position, I would never get promoted to that top NCO grade.

No problem. I said to myself. Sure, I was hurt by the entire unjustified relief situation, but there was nothing I could do about it. I mean, I knew there was something clandestine going on, but I didn't know what. So, I said forget it, and tried to put what happened in the past and move forward with my life. I knew that God had a hand in this and that sometimes things happen for a reason.

I didn't wonder about my future in the army at all. I just continued to do what I do. However, I have got to tell you that I never trusted my Company Commander from the moment he was assigned. But I never believed that a Lieutenant General in the U.S. Army would stoop so low as to deny a soldier a much-deserved promotion because of some kind of prejudice. Oh well, he did. Keep reading.

CHAPTER 34

Preparation for the United States Army Sergeants Major Academy (USASMA)

I had approximately six months before it was time to report to the Sergeants Major Academy. That period was a benefit to me, and I took 100 percent advantage of it. I spent lots of time with my family, focused on physical fitness, and also enrolled in a speed-reading course as suggested by the academy, just to name a few things that kept me occupied. I also spent lots of time trying to get the Nashville assignment out of my head. It didn't work.

Remember the General I worked for back at Fort Jackson? Well, here is the rest of the story. Every Christmas for several years, my wife and I received a Christmas card from the Guinn family. We had no idea who this family was, but it was apparent that they knew us. Therefore, we responded with a Christmas card as well.

As I sat at my desk one day reminiscing, I thought about the Guinn family. I called my wife and got the name and number. I called the number, and a man answered the phone. The conversation went something like this: "Hello, my name is Henry Wilson. How are you?"

Before I could say another word, the man said, "Did you say Henry Wilson?"

"Yes," I responded.

"Hold on, Henry. I have someone who wants to speak with you."

Of course, I was in total wonderment, but I just held the phone. I was in total shock when this lady came on the line.

"Hi, Henry, this is General Coleman's daughter, Mary. How are you?"

I was so happy to hear from her, but I must admit, I was just as shocked. We continued our conversation for, I really don't know how long. However, I do remember one thing she told me that really stuck with me: "You know, Henry," she paused, "everybody who knows me knows you. I really appreciate everything you did for me when you worked for my dad."

Man, was I shocked, but I really appreciated hearing that from Mary. After our conversation, I sat there and went back almost twenty years in my mind to those days. I thought to myself again, what did I do to cause these people to like me so much? Well, I said to myself, I was just being me. I did what was required of me and then some. As I continued reminiscing, I thought, *here I am sitting here because I was not allowed to complete a job that was probably one of the easiest jobs I ever had.* I thought, *This is a God thing. There is something down the road better for me.* I put all of that stuff behind me and continued preparing myself for the USASMA.

CHAPTER 35

The United States Army Sergeants Major Academy (USASMA)

After a trying few months as First Sergeant of the Nashville Recruiting Company, I spent several months completing odd jobs assigned to me by the Battalion leadership. I also spent much-needed time with my family and focused heavily on physical fitness. I was in excellent physical and mental condition by the time I arrived at the Academy. I spent lots of time in silent prayer, hoping to rid my mind of the discriminatory actions placed against me. I believed the command wanted a white person in the position. It worked. God saw me through the rest of my career by enabling me to continue doing what I do.

I departed Nashville in early January 1988 via I-40 to Amarillo, TX. I took I-27 out of Amarillo toward Lubbock, TX. After traveling about 1,300 miles, I arrived safely at Fort Bliss, TX. The trip was uneventful. When I travel alone, I do lots of mostly pleasant thinking.

In this case, I did lots of thinking about my family. I thought mostly about if and when they could visit me at the Academy. Don't get me wrong, I have to tell you that I spent lots of time thinking about the Academy, mostly about what to expect. Since everything I knew about the training was hearsay, I concluded that I really didn't know anything about the Academy except that the soldiers who were selected to attend this training were the top enlisted soldiers in the U.S. Army and a few from other branches.

One thing I heard that stuck with me was that the training was extremely hard. I mean, since the Sergeants Major Academy produced training for top-performing Senior Noncommissioned Officers, E8s and E9s, why wouldn't it be hard? I don't usually believe anything I can't confirm; however, I believed this one.

At one point, as I traveled closer to my destination, I actually thought myself into a frightening frenzy. About the same time that I was feeling the stress of what I thought was to come, I hit a pheasant that flew across in front of my van. I pulled to the side of the road to check out the damage, if there was any.

I concluded that this was a sign from God for me to get control of my mind. Since there was no damage to my van to speak of, I just walked around for a while. I thought a lot about the Nashville assignment. Situations that happened there, along with my perceptions, kept popping into my mind, no matter how hard I tried to forget them.

You know how scammers and dishonest people hack into your computer? Well, it seemed like the devil kept hacking into my mind about the Nashville assignment. I have got to tell you that I have purchased an excellent security system that never fails to block the devil's every move to try to keep me down.

God was consistent in guiding me toward the answers that I needed to keep me focused on the right things. I rested as I communed with God and thought about my family. This really helped me calm my nerves. I said to myself, "It is what it is, Henry."

I pulled back onto the road and continued to El Paso, TX. Then it happened. As I thought about all of those "top-of-the-line soldiers," "the crust of the Noncommissioned Officer ranks," "the super soldiers," it hit me. This voice in my head came to me once again. Just as plain as day, the voice (God) said, "Henry, Henry, listen to me. You are one of those soldiers." That's all the voice said: "Henry, Henry, listen to me. You are one of those soldiers." At that moment, I could feel a big smile coming over my face. I tell you no lie—from that point, it seemed that I couldn't wait to get to Fort Bliss.

I realized that you don't get selected for this prestigious assignment unless you are at the top of the Noncommissioned Officer ranks morally, physically, and your performance is beyond reproach. You must be exemplary in every way to attend this Academy. I still wondered about the processes of the Academy, our living arrangements, and a few other things, but there was nothing that I was concerned about.

I remained calm and relaxed as I proceeded to my destination. The changes in the landscape were very captivating. My mind kept going back to the big differences in nature from Tennessee to Texas and what it would look like near the Mexican border. Oh well, I guess I would find out.

With all those thoughts bouncing through my head, my family never left my mind. Then there it was, the campus of the Academy. As I pulled onto the Academy grounds, I really admired the design of the campus from my vantage point. It was absolutely beautiful. Still just a little nervous because of not knowing what to expect, I proceeded to the check-in point.

In-processing was very organized and professional. We were assigned our class and group numbers and living quarters. I have attended many basic, mid-level, and senior courses during my service in the U.S. Army. This senior leadership training would undoubtedly be the last leadership training I would attend during my career. What it did for me was send my mind back to the first Army training I attended back in 1966: Basic Training. Man, what a difference.

Back in September 1966, I remembered riding a big green bus onto a Basic Training Battalion area with Drill Sergeants screaming and spitting tobacco juice all over you. They were jumping up and down, cursing, and calling us all sorts of derogatory names. I actually thought they were crazy. Now this atmosphere was totally different. It was a sign of the changes that soldiers witness as they attend the many different types of training over the years.

This senior leadership training was indicative of where we were in our military lives. If I were to compare those in-processing procedures to my college in-processing procedures, the USASMA would be at the top of the list for organization and accuracy. We continued our in-processing with a simple physical, which included a blood pressure check and other examinations.

We were assigned quarters on base; however, we had the option to live off base if we desired. Some soldiers did live off base, but not me. I am a simple person. I took what they assigned to me, which turned out to be a very suitable one-room apartment with my own health maintenance facility. What they did was take some old World War II barracks and turn them into nice one-room apartments.

The only drawback was that some soldiers had to share a bathroom with one other soldier. The bathroom was built between and connected to the two one-room apartments. I guess I received the private bath because of time in grade or just the luck of the draw. Of course, I brought in my own refrigerator, coffee pot, and small grill. With that, I was set.

We were given time to check into our apartments, which also included cleaning teams. I also added a telephone so that my wife and I could continue to manage our home affairs and raise our children together while we were separated. I arranged my furniture to my satisfaction, set up my coffee pot, crockpot, TV, and other items. My friend from Fort Campbell, KY, had a crockpot as well, and that made me well set for food prep for the next six months. I was set, but I still had the jitters in my gut because of the unknown. I remembered thinking, It is what it is.

CHAPTER 36

Class Starts

Class started at 8:00 a.m. We met our instructor, who introduced himself and proceeded to give a synopsis of his military and personal life. After he completed his introduction, each student was asked to do the same. It was very interesting hearing how similar we all were in our military lives, except I seemed to have slightly more experience in the number of career fields than most. We seemed to show some differences in our personal lives. Our training manual was one thick sucker. Once we were given our daily and weekly instructions, we were off and running.

Of the three subject matter segments, my group started with the Resource Management Training Segment. Since I had spent about 20 years in leadership to this point, I was very versed in this subject. However, I still had the jitters. (Smile.) Once we got started with the training, it was over. Man, I smoked that training. (Smile.) Let me explain.

Reading was always my least favorite thing to do, except when it came to my work. Reading the Primary Regulations and Standard Operating Procedures pertaining to my work was always a priority for me. However, I still had the jitters until we got started with the class. Each student was charged with reading a segment of the manual orally, starting with the first chapter. The students, collectively, would then discuss what was read. Of course, the reader had the privilege of kicking off the discussion if desired. I got to tell you, I surprised myself with the knowledge I had of the subject matter.

You see, at this point in my career, I had been assigned to a leadership position for most of my twenty-one and a half years of service. What helped me gain so much knowledge in the Resource Management area is that during these assignments, I paid attention to anything and everything that touched my job responsibilities in any way. I paid particular attention to my most important resource, my employees, and as I mentioned earlier, I studied as many Army Regulations and Training Manuals related to my job as I could. In addition, I always tried to find better ways to get the job done. Therefore, shortly after we started this segment of training, the jitters were gone.

I really enjoyed the reading sections of each session, but let me tell you something. I was extremely vocal. I just could not hold back my knowledge of the subject. I talked so much that sometimes I had to hold back to give other classmates a chance to participate. If no one spoke up, I took the opportunity to add my knowledge on the appropriate point in the subject.

The Resource Management Section lasted a little more than a month and a half, as did the Leadership and Military Science Sessions. The Leadership Session went about the same as the Resource Management Session. As I mentioned earlier, at this point in my career, I had served twenty of my twenty-two years in a leadership position. Therefore, I had a world of knowledge to share in this area as well. The Military Science Session was slightly different, but I held my own. Keep reading.

The amount of participation the students gave in the Leadership phase of training over the Resource Management phase did not surprise me. I know that most, if not all, Senior Noncommissioned Officers have a substantial amount of leadership experience. One problem, as I see it, is that a lot of us, as leaders, tend to forget the collateral things we need to complete our mission. This is where I stood out in this segment of the Academy. Let me explain.

An assigned mission of a section is only a part of a unit's, Company, Battalion, Brigade, etc., overall mission. That is why, as soon as I could, during each of my assignments, I would try to make contact with other section leaders. In my own way, I always tried to find out a little about their assigned mission.

I have the distinction of having served in at least five career fields with several different positions to my credit in each career field. This played a big part in giving me a substantial amount of leadership experience. This experience helped me realize that a leader must pay attention to everything he or she needs to accomplish the assigned mission. All of us know that our employees are our most important resource.

We assign the best soldier to the right job, and you must hold those soldiers accountable. For example, the technical and supply sections, just to name a couple, can make or break a unit. I always stayed in touch with my section chiefs through scheduled briefings and unscheduled communication. My intent was never to spy on them. What was important to me was to find out where they were with the progression of their section responsibilities and to lend a hand if needed. These types of checks and balances let me know where we were as a unit, as it related to the success of my overall mission.

In addition, to keep each section in line with my completion timeline, I always gave constructive feedback and guidance as needed. I made sure what they were doing was in line with the successful completion of my assigned mission. Lastly, but surely not the bottom line, I made sure I stayed in touch with my counterparts in other sections within the unit. In addition, I always identified employees who possessed the necessary qualities to move into a supervisory role at a moment's notice. Ok, back to the Leadership Training.

I gave just as much, maybe more, constructive information to my fellow students during the Leadership phase of training. Let me share a couple of experiences I had during this phase. One student read a segment from the manual, which had something to do with First Sergeant's responsibilities. When he began to give his interpretation of what he read, this is a close example of how the conversation went.

The student said, "First, let me give you an example of the United States Army First Sergeants. The Recruiting First Sergeants are not the same as First Sergeants in other career fields because their responsibilities are different."

I interrupted him before he could go any further. "Hold up!" I responded.

"What do you mean that a Recruiting First Sergeant has a different responsibility?"

Without hesitating, I continued, "A First Sergeant is a First Sergeant. Their responsibility is no different. They have the responsibility of accountability, training, and welfare of all NCOs assigned to their unit, just to name a few, and to report the same to the Company Commander. Their Company demographics and mission may be different," I said, "but their responsibility to help the unit complete a successful mission is no different."

I apologized to the student and my fellow classmates for the interruption and for being so blunt. However, he and the other students could tell that I was very passionate about this issue.

You see, in addition to reading the training manuals on First Sergeant's duties, I paid attention to First Sergeants as they went about their work whenever I was in their presence. A leader in any profession should not have tunnel vision when going about their responsibilities. They should pay attention to everything that's going on around them because it could affect their mission. A true leader should never say, "That's not my job," because you can always learn from other leaders' responsibilities. The fact is, there is always a good possibility that you may have to step into that role on a moment's notice.

Quick story: Our oldest sister sent our oldest brother and me to the corner store about a mile from our house. Since my brother was the oldest, she handed him the keys to her standard-shift vehicle. Neither of us had driven a car before. The area we used to get onto and off the road from our house was lower than the actual road.

My brother tried four or five times to get onto the road, so he gave up and took the keys back to our sister. As he handed our sister the keys, I told her that I could drive the car. Without saying a word, she handed me the keys. I drove the car up the small embankment with no problem, drove to the store and back without any problems.

What's my point? I learned to drive a car without ever actually driving a car. Each time I rode in a car, I paid attention to everything that the driver did as they drove. This included what they did to start the car, what they did to shift the gears, how they steered the car, and what they did when the car didn't have a turn signal switch.

As I mentioned earlier, I used that quality in every position I held before, during, and after my military life. That is why I was able to counter the student's perception of non-combat First Sergeants so easily. The student didn't say another word. Oh well. The training continued. I continued my very well-received and very helpful input during this section.

I realized that I was talking a lot, but I didn't hesitate to continue when the need arose. I knew that the information I shared was based on my extensive experience, but I never considered how much other students received it until one student approached me during a break. About halfway through the Leadership Phase of training, a Sergeant Major, get it, a Sergeant Major who is one grade my senior, approached me. With a look of amazement on his face, he asked, "Henry," he said, "how the hell do you know all that stuff?"

Man, was I pleasantly surprised. I really don't remember what my response was exactly, but based on his reaction, it must have been on target. "Stupendous," he said. "You are downright stupendous." He repeated it as he walked away to take a smoke.

To be honest, I really didn't know what the word "stupendous" meant; however, because of the way the word sounded, I went and looked it up right away. (Smile.) After learning the meaning of stupendous, I felt absolutely great, and I extended my vocabulary. (Smile.) But I stayed grounded. I continued what I had been doing during each phase of the Academy training.

CHAPTER 37

The Military
Science Phase

The third and final phase of training for my group was the Military Science Phase. The Military Science Training dealt primarily with combat-type situations. Of course, Leadership and Resource Management are always embedded within any unit, no matter what mission is assigned. Based on my estimation, Combat Arms students made up over half of the class total.

The only combat experience I had at this point was Basic Training, Drill Sergeant Training, and Drill Sergeant Duty, where I trained new soldiers in basic combat situations. Of course, I had a little combat experience when I served in Vietnam. Don't get me wrong, what little combat experience I had at that point could never compare to that of a trained combat soldier. This also includes those support soldiers who are assigned to combat units.

As I mentioned, many of the soldiers in my Military Science Phase of training seemed to be combat soldiers. I looked forward to interacting with them. I had no idea what was going to be taught, but I was excited for what I thought I would learn in this phase. As it turned out, I learned a lot about the makeup of a combat division.

As you may imagine, I did a lot less talking and a lot more listening while in this class. The instructors had maps depicting different combat-type situations posted throughout the classroom. I participated in lots of written combat situations that I imagined happened during an actual war or conflict in which American soldiers were involved. I was particularly interested in the discussions by soldiers who made up different segments of a combat company, battalion, brigade, and division.

They talked about how different elements of a division support the main element, which is the infantry soldier. They talked about how commanders and platoon leaders would call in the artillery and air strikes in support of the infantry in combat situations. I knew that the Combat Engineers, the Military Police, and the Cavalry, for example, all existed within each combat division. I hoped to learn how those soldiers helped to support the division and when they were needed.

I was very impressed with the passion a lot of the soldiers showed as they discussed their combat experiences. I had no idea that the knowledge I received during this phase of training would help me in my next assignment, but it did. I'll explain later. Keep reading. I'll just say that I am so glad I was totally focused during this particular phase of the academy training.

Our training during the course included several orientations by specialized professionals from different companies and from within our United States Military as well. We were involved in different functions that we would most likely be involved in during our next assignments. For example, we were fortunate to attend a full Formal Military Dinner.

The dress for this function was Formal Military Attire, which included, at a minimum, our dress blue uniform with the bow tie. I had attended Formal Military Dinners in the past, but this one, without a doubt, was a new experience, and I loved it. I was fortunate to have been selected to shout out a salute to fallen soldiers at the appropriate time. The salute that I shouted out was one of many salutes to soldiers and veterans who are worthy of this type of notoriety.

The most exciting experience for me was getting the opportunity to sit across from two Sergeant Majors who had received the prestigious Medal of Honor for their service in Vietnam. I know they felt me staring at them from across the table. I'm sorry, but I just couldn't help myself. One of the Medal of Honor winners' wives accompanied him during the formal affair.

I could not help but notice her emotions during the program. I mean, I could not imagine what she had gone through when her husband was a Prisoner of War and up to the point of the dinner. This soldier was home after being a POW for seven years. As I read his heroic actions prior to his capture, along with one other soldier of his unit, I actually felt chills rise over my body as I fought back my tears. I really admired both of those Sergeants Major for what they had done for the

United States of America, the United States Army, and their comrades. I would really hate to have been in the situation each of those soldiers faced. I could really talk about these guys for hours, but I will leave it at that.

Another perk of this course was that each student had the opportunity to take college courses during the program. I chose to take Business Law and Business Math, which gave me six semester hours of college credit. I participated in different types of extracurricular activities, which really helped clear my mind from all the studying I had to do. Other community activities included washing cars, bowling, organized physical fitness, and Armed Forces Day activities. This school was really organized to cover a broad spectrum of the mental and physical requirements and needs of a senior NCO.

This course was truly amazing. It helped me tremendously. In my opinion, the school's architectural design and curriculum were well developed with the students in mind. I really enjoyed every part of the course.

CHAPTER 38

Graduation and Preparation for My Next Assignment

After about six months of outstanding training, it was time for me to move on to another assignment. I must say, I graduated with a superior rating in Leadership and Resource Management and a successful rating in Military Science. I looked forward to my next assignment, wherever it would take me.

I received a call from one of my previous Battalion Commanders and his Battalion Sergeant Major. They offered me a job as their Battalion Operations Sergeant Major. I recognized that the Sergeant Major was one of my counterparts who had been the First Sergeant of the Chattanooga Recruiting Company of the Nashville Battalion.

I really felt great that the Battalion Commander and Battalion Sergeant Major contacted me to fill the Operations Sergeant Major position. Of course, I had to inform them that I could no longer work in recruiting. Although I was informed that there was talk of my being considered to teach Leadership at the Academy, obviously, that didn't happen. I ended up at Fort Riley, Kansas.

CHAPTER 39

My Assignment at Fort Riley, Kansas

Ireceived a call from the Brigade Sergeant Major welcoming me to the United States Army Correctional Brigade. I was to become Sergeant Major of one of two Battalions within the Correctional Brigade. I felt it necessary to inform the Sergeant Major that, unfortunately, I was not a Sergeant Major. Subsequently, I became the Noncommissioned Officer in Charge of the Brigade's Operations, Plans, and Training (OP&T) Section.

Of course, the first order of business was spending time with my new boss, the Chief of Operations, Plans, and Training, who was a female Major. I only mention this because it was the first time I had a female as my immediate supervisor, and I had absolutely no problem with that. As always, I did what was necessary to get the job done. I want to point out that once I got into the crux of my responsibilities, my boss supported my decisions. She never insisted that I change any of my methods of operation. That alone was a win for the organization. I'll explain as I continue. Keep reading.

As the Major continued to brief me on our mission, my first impression of her as a person and as a leader was very positive. As we continued the briefing, she spoke to me as the senior Noncommissioned Officer in the section. What was especially noteworthy to me was that the Major didn't do anything to force her position or grade on me. The key to my successful relationship with my boss was that she recognized my abilities and allowed me to do my job. When I needed her approval for an action, I went to her, and she supported the decisions I made to get the job done. In addition, I always kept her totally informed as to what was going on in the Operations, Plans, and Training Section.

Immediately, I planned a briefing from each of my section chiefs during the first week of my assignment. It was ironic that an opportunity arose that showed me that my new boss was like other commanders I had worked for before my Nashville assignment. I will explain. Keep reading.

As my boss and I talked during the briefing, my mind kept reflecting on the Nashville Recruiting Company and Battalion Commanding Officer, and other commanders up the chain. I thought to myself, *Henry, you have never had a problem with your superiors during your entire military career. You must realize that Nashville was different from any assignment you have had for 22 years. You still don't know what really happened or why they treated you like you didn't belong there from the start. You can't let what happened there, whatever the reason was, affect your performance. Do what you do and shake that mess out of your mind.*

Within the first day or so, my boss not only introduced me to our staff, civilian and military, but she also introduced me to our Command Staff as well. Our Correctional Staff consisted of many very knowledgeable military and civilian personnel. The Brigade had the responsibility of managing the custody and control of up to six hundred prisoners. The Brigade consisted of two Battalions and a Headquarters Company. The prisoners were soldiers who had been convicted and adjudged by action of a Court Martial conducted according to the Uniform Code of Military Justice.

One Battalion housed Medium Custody Prisoners who were restricted to the Battalion area encircled with a wire fence. The other Battalion housed Minimum Custody Prisoners who were given the opportunity to work in our Vocational Educational Training (VET) Program. This program was established to help prepare those future civilians with skills they could use once they were discharged. This is where our staff came in.

The VET Program was managed by a Senior Noncommissioned Officer with an assigned GS7 Technical Support Specialist. Although I had no control over the civilian GS7, the program came under my supervision. Thirty to forty VET sites had been established throughout Fort Riley. These sites included Small Engine Repair, Dental Tech, Golf Course Maintenance, and Food Service, just to name a few. Within our Brigade, we assigned prisoners to the Barber Shop, Building Maintenance, Woodworking, the Garden Center, and others.

We were responsible for assigning the Minimum Custody Prisoners to all VET sites. I'll get into some details about these sites shortly.

I planned to visit the VET sites within the Brigade first. I felt that what was going on within the Brigade could be a reflection of what was happening on VET sites throughout the main post. As I visited each site, I received a briefing conducted by the NCO Site Chief. Two sites, the Barber Shop and Building Maintenance Sites, also had civilian technical support. I mentioned technical support because it was important to know how I would proceed with what I had to do to fix problems that I identified within those areas.

As the Site Chiefs conducted their briefings, my eyes continued to scan their areas of responsibility. During the briefing, I moved around the VET site, noting the good things and things I had questions about. At this point, I had no intent to fix anything. I just wanted to take in as much information as I could about the VET Program, as well as every element of the Brigade. Although the program was in place, I found some things that needed to be addressed. The briefing by the VET Program Chief and the Technical Advisor gave me the opportunity to make my first decision, which was a no-brainer. Let me explain.

The first problem brought to me by my Section Chief was with two prisoners who worked in the Barber Shop. Toward the end of the briefing, the Section Chief stated, "Top, I have a problem in our Barber Shop." I replied, "Okay, what's up?"

The Chief stated, "Two of our best barbers had a disagreement and started fighting in the shop."

Without hesitation, I stated, "Move them."

The Chief said, "But they are our best barbers. We use them to train the other prisoners." I thought to myself, there's another problem, but I'll address that at the appropriate time. Without hesitation and with eye-to-eye contact with the Chief, I said, "Obviously, you didn't hear me, Sergeant First Class. Move them! And I want you to move them right away. One more thing, Sergeant. I want you to make sure you demote them back to Medium Custody."

"Okay, Top," he replied. I thought that was the end of the story, but guess what, the subject popped up again, which turned out to be a good thing for me. Keep reading. I will continue this story shortly.

I took the opportunity to observe the prisoners going to their various work sites, which took place each weekday and for some prisoners on the weekend as well. Those who worked at the various sites within the Brigade walked to work, and the rest of the prisoners were transported by someone from each VET site to the main post.

What disturbed me was, once the prisoners left their company area, there was absolutely no prisoner accountability. Other than knowing which VET site the prisoner went to, we had no way of knowing which person picked up the prisoners, and if they were on the VET site during the work period. This was perhaps the most severe deficiency I had seen during my observation of the operations of the programs under my supervision. I knew this deficiency could cause the Brigade big problems.

Therefore, I started addressing the issue right away. I familiarized myself with the regulations and procedures that were in place for the Brigade. But first, I felt it was best for me to continue visiting each VET site on and outside of the Brigade so that I could get a complete picture of what needed to be done.

Shortly after the briefing by the VET Chief, a group of us stood in front of my desk just talking. The group included the VET Chief and the VET Technical Advisor (GS7), the Operations Officer, who was a Captain, and, of course, me. Suddenly, my boss, Commander of OP&T, approached us. As she walked toward us, the VET Chief stated to her, "Ma'am, we have two prisoners in the Barber Shop that got in trouble, and we want—"

At that point, I exploded. "Hold up!" I shouted in a very loud and demanding voice. "What the fuck did I tell you to do, Sergeant First Class?" At that point, I didn't wait for him to answer. With eye-to-eye contact with the Section Chief and just inside his physical space, I continued, "That's what the fuck I mean for you to do. Now!" The Major never stopped walking. She threw her hands up, turned around, and left. At that point, I didn't stop to think about what her actions did for me.

I took the opportunity to give a little information about what the Section Chief should have done and why. I continued, "That was an issue you should have fixed without my or the Major's involvement. When a prisoner violates any of our policies on a VET site, he or she will be removed from the site without hesitation. Do you understand me?"

"Yes, Top," he responded. I then addressed everybody in the group. "Do y'all think I'm being too demanding?"

"I don't think you are being too demanding, Top. I agree with you," the Operations Officer stated.

"Thank you, sir," I said, and went on to say to the Section Chief, "Listen to me, Sergeant, while we are on the subject, you will never have a prisoner train another prisoner. That's why the GS7 was hired. He is there to teach the prisoners to cut hair. Make sure you fix that on each of our VET sites right away."

I went directly to my boss and apologized for my language, and I explained to her what the issue was. She agreed with me. I thanked her and continued with my work. My boss's actions spoke volumes about the type of support I could expect from her as long as I was making good decisions.

As I visited each of the VET sites, I found some deficiencies that, in my opinion, needed to be fixed as soon as possible. I felt that there was no cadre supervision at the VET sites on the main post. As I talked to VET Site Program Supervisors and soldiers under my supervision, I heard chatter about drug involvement on some sites. I also noticed a disproportionate number of Black and Brown prisoners compared to white prisoners assigned to the more desirable VET sites than others.

For example, there were a large number of Black prisoners assigned to inside jobs like the clinics, the Barber Shop, admin jobs, etc. You see, over my lifetime, I have been a victim of discrimination based on race and gender. I vowed that I would never discriminate or allow anyone who worked under my supervision to discriminate against anyone for any reason.

As soon as I confirmed what was obvious, I jumped on this problem. In private, I spoke with my VET Chief, who made the assignments. First, I explained to him the deficiencies I had seen during my site visits. Then I asked if he had any particular status information on the prisoners that aided him in making these assignments. He gave his explanation, which was questionable to me.

Out of respect, I told the Chief that I heard and understood what he was saying. However, I needed him to make some adjustments to his assignment procedures right away. At this point, I did not share this information with my boss. My intent was to get this, what I considered a very critical problem for the Brigade, under control before I briefed my boss.

I learned over the years that some things that you encounter on the job are NCO business. What you should do is fix the problem, then brief your boss, letting them know that you uncovered procedures that were inconsistent with Army Regulation and what you did to correct the problem. This way, you can take total responsibility if the boss disagrees with your actions. Of course, I had no doubt that what I had done was the right thing and that my boss would approve because it was within regulations and was the right thing to do. I also knew that the VET Chief would fix the problem.

As I continued my counseling with the VET Chief, I first wanted to explain to him the exact deficiencies that I had found in his program, after which he wouldn't have any doubt as to what I was directing him to correct. I had learned during the short time of my assignment that I could, once I got his attention, always depend on him to do his job well. He never wavered when there were necessary changes put into place by our Brigade leadership, my boss, or me. My VET Chief had great potential, and with my guidance, he would become a very successful leader.

Anyway, as I continued my counseling session with the Chief, I shared with him the things I had noticed as I visited the VET sites as they related to assignments. I told the Chief, without hesitation, that I believed I knew what was going on and that I would not tolerate any type of discrimination in our VET site assignments or any program under my supervision. I detailed the deficiencies and explained that I wanted him to make correcting them his immediate priority. The Chief responded, "Ok, Top," and proceeded to fix the problem right away. Of course, I followed up just as I have always done and as any effective leader should do.

Sometime later, I completed the VET Prisoners Work Call Standard Operating Procedures (SOP). It took me longer than expected because there were so many details I had to check and double-check as they related to what was in place and what needed to be changed. In addition, I discovered some serious problems within the Building Maintenance Section that needed to be fixed. I'll talk more about that later. It so happened that the strangest thing occurred just as I asked the Operations Officer, who was acting for my boss, to look over the document and give me his input. The document was simply Standard Operating Procedures pertaining to the custody and control of prisoners while they were on a VET work site.

Just as the captain started reading the document, we got word that a prisoner was missing from his VET site on the main post. Of course, we activated the prisoner search procedure and quickly located the missing prisoner at a house in the town adjacent to the base. I, along with the Operations Officer, called the Battalion Commander for an appointment. Since the commander was responsible for the escaped prisoner, we wanted to present the after-action report directly to her. We were granted the appointment and met with the Commander right away. During the meeting, we took the opportunity to introduce the recommended changes that we asked her to put into place.

After reading the after-action report, the Commander read the recommended document right away and asked, "Why isn't this in place already?"

"Well, ma'am," I said, "when I was assigned some time back, I discovered that there were no effective work call procedures in place for prisoner accountability. I thought we needed to have SOPs that were more in keeping with the Army Regulations governing confinement facilities and custody and control of prisoners. I asked my boss to look them over so that I could make the necessary changes before presenting them to you. The fact that all VET prisoners are assigned to your Battalion, we wanted to get your approval to put these changes into effect. I want to point out that your Drill Sergeants will be heavily impacted as well."

"Do it," she replied.

We moved forward with the plan immediately. I assisted the VET Supervisor in drawing up the necessary forms that would be used by each VET Site Supervisor to account for their assigned prisoners. I briefed my OPT Staff, and I also briefed the Battalion personnel who managed the VET prisoners when they were off duty from their VET Sites. After briefing all necessary personnel, we moved forward with the plan. Of course, we had pushback from several soldiers. My message to them was that the plan is in place. I continued, "We will make changes, if necessary, only after you have followed the plan as written. Remember, you must work the plan. Do not let the plan work you."

The SOP dictated that the Battalion would hold a work call formation each weekday morning at a designated time. A representative from each VET site would sign for their prisoner(s) and transport the prisoner(s) to and from the VET site. A copy of the signed document would be maintained by their VET Site Chief. My observation of the changes I put in place during the next few weeks showed that the program worked perfectly, and everybody was happy.

As I visited VET sites within the Brigade, I purposely walked through each company area. I thought this was the best way to learn the layout of the Brigade infrastructure. As I walked through, I noticed a lot of small projects being built in different company areas. I noted in my mind that those were also projects that required many different types of building equipment and supplies. During the briefing by the Building Maintenance Supervisor, a Senior Noncommissioned Officer, and his Technical Advisor, a General Service Employee, I asked lots of questions. Of course, my eyes had already told me plenty as I walked throughout the Brigade and within the Building Maintenance Workshop. I just wanted confirmation of my findings.

I wanted to know what kind of supplies and equipment they had, how much of each they had, and how they ordered them. I also wanted to know how they issued it to the customers and, most importantly, how they accounted for the supplies and equipment they issued or loaned to each person or unit. I have to tell you that I was really shocked when I found out that there was absolutely no audit trail for the building supplies that were used or given to Brigade personnel. I mean, absolutely no accountability.

As I toured the facility, I estimated thousands of dollars' worth of building machinery that was used within the shop. Don't get me wrong, I don't claim to know what each piece of equipment was used for. Nevertheless, it was there, and the VET prisoners were being taught what the machinery was used for and how to operate each piece of equipment.

This was absolutely a good thing. Training of the VET prisoners was our business and was perhaps the only positive thing I found in the shop at this point. In addition to the equipment, I estimated thousands of dollars in supplies in the shop as well. "How do you account for the supplies that you issue to the customers, Sergeant?"

I asked. "We just give them what they want, Top," the Site Supervisor responded.

"Are you telling me that you don't have them sign for what you issue them? Do you know what they are using the supplies for?" I continued. "All we know is that they use it to build stuff in their company area." Both the Site Supervisor and Technical Advisor agreed.

With disbelief showing on my face and in my body language, I told them that the procedures they were using, or lack of, had to change. "Guys, listen to me. Anytime you use anything paid for by the government, you must take total responsibility for its use and accountability. If you don't have an audit trail, it will, without a doubt, come back to bite you on your ass."

I directed them to halt issuing anything from the shop until I got back with them. Shortly after that meeting, I drew up a plan for the shop personnel to initiate accountability for all supplies and equipment requested by authorized personnel. The plan required the customer to complete a request form listing the supplies or equipment they were requesting and their purpose for the items. It also had a return date for borrowed building equipment. The request was to be submitted through my office for approval. I kept a copy of the approved request and forwarded it to the shop to be filled. Once I briefed my boss on the new Building Maintenance Request Procedures and got approval, I put the plan into operation.

Of course, we got pushback because of the change. Some said that I was trying to keep them from getting building supplies. Some asked the question, "Why the change?" They felt that everything was working just fine. I was very open and direct about the change.

I explained that the change was necessary only to facilitate accountability for the supplies and equipment that were provided by the U.S. Government. I went on to explain that we would be accountable for anything that was provided by the U.S. Government. It is inevitable that someone, somewhere, has the responsibility to account for all the money spent on every piece of equipment and supplies that we use in support of our mission. Believe me, someone is tracking this stuff.

I went on to assure our customers that I believed that there would be an audit at some point in the future. As I closed the briefing, I told them that whether there is or is not an audit, we are going to do the right thing. Guess what, just a few months later, we received a visit from a member of an audit unit from, I assumed, the main post. I'll explain later. Keep reading.

I also assured them that every customer would continue to get the supplies and equipment they requested. The only change is that your request will be in writing and will be forwarded through my office. The request will be approved and forwarded to the Building Maintenance Shop within twelve hours.

The much-needed change worked well. From the date I established the change, we could account for every piece of building material and supplies, and equipment that was issued by the Building Maintenance Shop. Don't get me wrong, I know that there were some materials, and especially supplies, that slipped through the cracks. However, my boss and I, and our customers, were very pleased with the change.

Our Woodworking, Landscaping, and Garden VET Sites were working extremely well. The Woodworking and Garden VET Sites sold items they made or produced in our Thrift Store. The only concern I had about the store was the management of the funds it brought in. After following up on the operations over a period, I determined that everything was okay.

The Thrift Store was open to all military personnel and their families. I supported each VET Site, including the Barber Shop, by frequently utilizing them. The prisoners who worked in the Woodworking Shop made very durable and beautiful picture frames, different-sized trunks, and other items. I still have items today that I purchased from them, including a small trunk that I gave to my mother-in-law for Christmas of 1989. I also have a family photo encased in a unique picture frame made by the prisoners.

As I mentioned earlier, we had great employees, both military and civilians, who knew their jobs very well and were very diligent in getting the job done. I noticed, however, that there was no Training NCO assigned to our section. The Training Noncommissioned Officer (NCO) would manage training records for all enlisted soldiers who are assigned to our section. This NCO would also recommend deserving enlisted soldiers to attend training that's available for their grade. He/she will also be responsible for recommending any qualified enlisted soldier to be assigned to our section. I spoke with my boss about the need for someone to manage our enlisted personnel training program. After she gave me the thumbs up, I moved quickly to assign a Training NCO. I also spoke with the Brigade Sergeant Major about assigning personnel to our Operations, Plans, and Training Section, which I will get to shortly.

I believe that every Senior NCO should get as much experience in as many positions as possible when the opportunity arises. Our section had very capable Senior NCOs; therefore, I selected one NCO, a Sergeant First Class (SFC), who had the grade and, I hoped, with my assistance, could do the job. Of course, I went over the position responsibilities and our expectations for the position with him.

Because he had not held the position before, I assisted him with finding training manuals and regulations that he would need to familiarize himself with in order to do the job effectively. During the next few months following his assignment, I held training sessions with the SFC and made sure I was available to assist him when needed. I was sure he would do a good job.

In the meantime, I met with the Brigade Sergeant Major to discuss Correctional Specialist assignments for our section. What had been the practice for assigning soldiers to the Operations Plans and Training Section was that the Brigade Sergeant Major would charge the Battalions with sending soldiers from their Companies to work in our section. During a few months after my assignment, I found out that newly assigned soldiers to the Brigade were being assigned directly to the Companies within the two Battalions. The Companies would, in turn, send soldiers from their Company to my section, the Operations Plans and Training Section.

Often, the Company would use this opportunity to get rid of their problem soldier(s), their least desirable soldier(s), if you will. That experience is what prompted me to request a meeting with the Brigade Sergeant Major to voice my concerns about the present procedures. During the meeting, I shared the following experience I had with one Company.

I recall a phone conversation I had with a First Sergeant who was tasked with sending me a soldier. "Hello, Top, I'm sending you a soldier," the First Sergeant stated.

"Thanks," I responded.

The First Sergeant continued, "He will be great for your section."

During my interview with the soldier, I found out that he was a problem child. In addition, he had a pending punishment imposed under the Uniform Code of Military Justice. No transfer action should have been taken on this soldier until after his punishment under the Uniform Code of Military Justice had been finalized.

I called the First Sergeant and informed him that I had directed the soldier to report back to him. I also explained why I would not accept the soldier. I asked the First Sergeant to send me a soldier that he would want to work for him, because that's the soldier who may be good for my section. I emphasized "may be."

I could tell the First Sergeant didn't like what I did or said, but no problem, just give me an effective soldier, and everything will be just fine.

During my meeting with the Brigade Sergeant Major, my major concern was that Correctional Specialists who were assigned to my section be treated no differently than the Correctional Specialists assigned to the Companies. I requested that my section receive soldiers from among those who were newly assigned to the Brigade, just like the Companies. The Brigade Sergeant Major approved my request and made it effective immediately. I thought that the urgency the Sergeant

Major placed on my request was simply great. It also told me that the things I had done after being assigned were well accepted by the Brigade staff. Of course, I felt great about all of the changes I put in place, but it did not affect my actions in any way. I kept doing what I do, which was keeping our mission as my priority in support of the United States Army.

About a year or so after I was assigned as Noncommissioned Officer in Charge of Operations Plans and Training, I realized that the Brigade personnel as a whole did not really know what functions the Operations Plans and Training provided for the Brigade. Our section was one of a kind, established for the custody and control of prisoners and to prepare United States Army soldiers who had been adjudged according to the laws under the Uniform Code of Military Justice (UCMJ) to return to civilian life. Therefore, those soldiers who had been trained and who had worked as a Correctional Specialist did not have a really good idea of what goes on outside of the normal prisoner control type situations.

Quick story: During an acting assignment for the Brigade Sergeant Major, I hosted a meeting with some of the Senior Correctional Drill Sergeants. During the meeting, one of the Drill Sergeants had the nerve to state that he didn't understand how I, having not been in their position, could tell them as Correctional Specialists how to do their job. Understanding his ignorance of my total Army experiences and his perception of my understanding of the totally changed correctional system, I simply stated, "Sergeant, I know more about custody and control of United States Army prisoners than you will learn during the rest of your career. I also know what you do and how you do it simply through observation."

I went on to say, "You don't know what positions I have held during my career; therefore, before you make those types of ignorant and unsubstantiated statements, I recommend that you do some research to get the facts straight. Research, like talking with the person you have questions about. Doing that could save you some embarrassment in the future," and I left it at that and continued with my meeting.

Those kinds of perceptions and some other facts gave me an idea. I wanted to come up with something to inform the Brigade personnel of what the Operations Plans and Training responsibilities were. After mulling over what I wanted to do, I briefed my boss, who thought it was a great idea. She told me, "Top, do what you have to do and don't hesitate to let me know if you need my assistance for anything."

"Thanks, ma'am," I responded with a smile and immediately started putting my plan into action. I briefed my Section Chiefs on my plan and directed them to put together an overview of their section responsibilities. After looking over each outline, I set a date for each of them to give me a complete briefing, at which time I made necessary changes to the briefings. I also made recommendations for each presentation. I sent invitations to key Brigade Noncommissioned Officers, and of course, I gave them the opportunity to invite other members of their unit. My extensive food service background gave me the opportunity to put together a few refreshments for the function.

Everything went extremely well. My Section Chiefs did an amazing job with their briefing. I received many positive, laudatory comments in reference to the orientation. Shortly after the orientation, I let my staff know that I was very pleased with their briefing and their work.

Remember the Building Maintenance Shop as it relates to accountability of equipment, supplies, and machinery? Well, it happened. An accountant from, I assume, the main post conducted an inquiry just prior to my retirement. He requested a meeting with the Section Chief and his Technical Advisor. Of course, I attended because, as you know, that section came under my supervision, and I would always support my people when needed.

The inventory sheets the inspector used contained lots of information as it relates to building supplies and equipment. I was not surprised to find that the numbers were outrageous for the period prior to my changing the shop distribution procedures.

The representative asked specifically about the supply statistics that showed very high numbers for supply usage. The Section Chief and Technical Advisor made comments that did not make sense to me. The representative then asked me if I knew why the numbers were so high. I responded that I had an idea, but I could not comment because I did not have the facts of what happened prior to my assignment. I remained in the meeting just as a support when needed. It was clear, however, that the information I gathered during the meeting confirmed my suspicion. There was a huge amount of building supply usage that could not be explained. However, I remained silent during the rest of the meeting. Oh, by the way, the Building Maintenance building, suspiciously, to me, burned down not long after that meeting. According to the fire department, the fire was officially caused by an electrical power surge. Yeah, right!

I was very pleased with the performance of the soldiers assigned to Operations Plans and Training. However, to me, there was still something missing. I mulled over possibilities for a while and decided that we needed some camaraderie and cohesiveness, if you will. In addition to our section's briefing on their responsibilities that we provided for our customers, I wanted all soldiers assigned to Operations Plans and Training to know each other. I wanted our families to socialize on and off the job.

To accomplish this, I put together a plan to have our soldiers get together during the afternoon of the last month of each quarter. This would include the lunch period and the rest of the afternoon, starting at 1:00 p.m., and would cover the rest of the work period. Soldiers were free to leave at the end of the work period. All soldiers were encouraged to participate; however, if anyone chose not to attend, they would be required to remain at work or take leave. Once they were given their assignment for the quarter, each section would be responsible for selecting the location, food, and recreation for the gathering.

After briefing my boss on the plan and a few questions for me, as expected, she approved the plan. I have to tell you, once my boss understood what it was, I was trying to do for our section and the Brigade, she was always supportive.

This initiative worked extremely well. Each of my Section Chiefs made sure their team accomplished their assigned responsibilities completely and in a timely manner.

Our team enjoyed plenty of great food, played games, and just communicated with each other during each recreation period. My boss, as I mentioned earlier, was always supportive. As a matter of fact, she hosted a function at her home. That was the last quarterly function I attended before I retired. It went extremely well. I received many positive comments about the things I had done for the benefit of our section, the Brigade, and the United States Army. I was absolutely pleased.

CHAPTER 40

The Unit Provisional Equipment Custodial Organization

As one of my additional responsibilities, I served as the Chief of the Unit Provisional Equipment Custodial Organization. We referred to the program by its acronym (UPECO). This program was a must for each Company within the 1st Infantry Division. The benefit to me was that it gave me the opportunity to continue learning more about our Combat Divisions. It also confirmed to me that some of the problems I had seen during my career continued to exist in some units in the United States Army.

One problem was that some of the Company's leadership did not completely hold Section Chiefs accountable for fulfilling their portion of the Brigade's overall mission. Leaders should always stay involved with their subordinates through training, inspections, briefings, and simple eyes-on. These are just a few ways that leaders can ensure that their Section Chiefs do their part in support of the unit's overall mission.

The mission of UPECO was to inspect the personal information of members of each unit to ensure that each soldier had a next of kin listed or someone to contact listed in their personnel file and verified by the soldier. This was extremely important for each soldier's family in the event of loss of life. All these items were to be checked by each member of my team once each month.

In addition, we checked Standard Operating Procedures (SOPs) for storing all residual items when the Brigade was deployed. This included ensuring that SOPs to secure each soldier's personal property were in place and completed in a timely manner.

Once each month at a predesignated time, I would gather my team of approximately 50 to 60 soldiers for a briefing and deployment to the 1st Infantry Division. Each member was assigned to a unit/company within the Division, including the Headquarters Company. Once I deployed the team, I would take my post in the War Room of the Division Headquarters.

Each team member would contact me once they had completed their inspection, unless they needed assistance, at which time they would contact me for guidance. Otherwise, they would complete the necessary forms, brief the Company leadership on their findings, and give them a copy of the inspection sheet. I would receive a copy of the inspection sheet once we returned to our unit. I would complete a master copy of my files and brief my boss immediately.

As fate would have it, the opportunity to use the information we had gathered for approximately two and a half years came unexpectedly. I was notified that the Division was being activated and would be shipping out soon. I was also given a date and a possible time that I was to brief the leadership of units from the Division Headquarters to each Company within the Division. I was very curious to find out if the findings from our inspections benefited the units as they prepared for deployment. I got to tell you, I was not optimistic because some of the same deficiencies were reported each month for some of the same Companies.

As I received questions after my briefing, it was obvious that some of the Company's leadership did not correct their company's deficiencies. One Colonel, I never knew his position, asked several questions pertaining to the content of our monthly inspections. After I answered his questions, he proceeded to ask those in the meeting why he had not received any information about these inspections.

He went on to say that my briefing perhaps provided the most important information that would protect the soldiers while they were deployed. He then directed the Company Commanders to provide him with the status of their Company as it related to each of the questions on our inspection forms. Needless to say, he gave them a very short suspense date.

I felt good about my briefing and my team's involvement in the Division's readiness for deployment. The Division deployed, I believe, sometime in mid-1990. The Commander of Operations Plans and Training was very pleased with our involvement. I felt confident that our Brigade Commander would receive a call from the Commander of the Combat Brigade or his representative.

CHAPTER 41

My Retirement

Shortly after the Division Readiness Briefing, I pondered retiring from the U.S. Army. This was the first time I had seriously thought about retiring. I had always believed that I would complete 30 years of successful service. However, being separated from my family for more years was not sitting well with me. I also found myself drinking more alcohol than normal late into the night and early morning on weekends. There were three situations that helped me decide that I needed to back off on the liquor. Let me explain.

I attended a promotion party that was held for soldiers assigned to the Confinement Brigade. I had intentions of riding back to my house with two soldiers. During the party, they became angry with each other. Therefore, under those circumstances, I thought it best not to get involved with them.

I thought *you only had a couple of drinks. It's only two miles. I can do this.* I got in my car and drove into the entrance to our brigade. Just as I turned onto the road leading to the gate, the MP stepped out of his MP Stand and held up his hand, motioning me to stop. I thought, *This is it.* I just held my breath as he stated, "I'm sorry, Top. They told me to check everybody that comes through the gate."

Just as he stopped talking, the MP waved me through the gate. Still holding my breath. I didn't say a word or make any gestures. With a sigh of relief, I thought, *That's a message.*

Our Brigade Commander ordered a 4-mile run once each quarter for all soldiers and those civilians who wanted to participate. At the end of each run, food service personnel served a pancake breakfast with all the fixings just for the participants. As I ate breakfast and talked with a few soldiers, one soldier asked if she could speak with me in private. She informed me that I reeked of alcohol. I replied to the soldier that I really appreciated her letting me know.

You see, I had gone a little longer with the liquor than I should have on Sunday night. The worst thing was that the Monday morning 4-mile run brought all the liquor to the surface of my body through sweat. After the soldier informed me of the odor of alcohol coming from my body, I went directly home and took what seemed like an hour-long hot shower.

Guess what happened next? When I got to my office, my secretary informed me that I had to sit in for my boss during a Brigade Staff Meeting. When I arrived, the Brigade Commander offered me a seat on his couch with other staff members. I felt like everybody in the room smelled nothing but alcohol on me. I was grateful that no one said anything.

Another incident was when I was informed at the last minute that the Base Commander was giving all soldiers a four-day weekend. I went home, had a drink, packed my bag, and then went to bed with the intent of leaving around midnight, which would have given me about five hours of sleep. The problem was that I tossed and turned for about an hour. So, I got up and hit the road.

I got on the main road leaving Fort Riley—still on base—and bam, there were the Military Police (MPs) who had set up a DWI checkpoint. I was inside the test track with no way to get out. I knew I didn't have much to drink, but I didn't know what the MPs were going to check.

"Good evening," one MP said as he approached my van. He asked for my driver's license and ID card and proceeded to explain what he wanted me to do. They had set up about a forty-meter track of zigzag cones for drivers to navigate through. The MP handed my driver's license and ID back and explained that if the MP at the end of the track didn't stop me, I should keep going to my destination. I said to myself, "Henry, that's the third close call. You need to make some adjustments to your drinking habits."

In addition to those situations, I really didn't think I could mentally handle the disappointment of being passed over for the Sergeant Major promotion again, especially not knowing why this was happening to me. Therefore, I submitted my retirement papers, which were approved for the end of the fiscal year of 1990. Of course, I informed my boss that I would be retiring and when.

I have to tell you, I was extremely proud to have worked under another professional, hard but fair, commissioned officer. The Major, my boss, saw to it that my team gave me an outstanding going-away celebration. It was great. I still have some of the gifts my soldiers gave me. This assignment was my last as a soldier in the United States Army, culminating in a well-earned and extremely successful military career.

SUMMARY

I am extremely proud to have served twenty-four successful years in the United States Army. The experience I gained during my service was simply matchless. I was, perhaps, one of a few soldiers who successfully served in five different career fields and over ten different positions, including Enlisted Aide for a Base Commander. What is especially noteworthy is that I spent twenty-two of those twenty-four years in a leadership position.

I took pride in leading and training soldiers to do the best job they could while they could. I encouraged each soldier to pass on to others what they had learned whenever the opportunity arose. I treated all my soldiers the same, no matter what their race, gender, or any other status was. I learned to read people by what they say, when they say it, and their body language while they say what they say.

I felt that it was important that each soldier be cross-trained enough to know and understand the overall mission for my team or any team to which they were assigned. I believe that it was essential to hold each Section Chief accountable for his or her portion of the mission through appropriate checks and balances. I ensured my soldiers knew that once I gave them a mission, or they had the mission implied by the position, I would always follow up with them for timely reports. My checks and balances would continue until the mission was accomplished. I believed that scheduled reports and drop-ins were a must for the success of the mission.

I was successful in each position I held because I researched before and during my assignments, not just to change the procedures if needed, but first to ensure that the procedures were in accordance with Army regulations, and then to determine if the mission could be done better with well-researched, adjusted procedures.

During my lifetime I have been discriminated against because I am considered Negro, because I am considered Colored, because I am considered Afro-American, because I am considered African American, because I am Black, and because I AM!! In addition, I was denied promotions in a Federal Government position because I am a retired U.S. Army veteran and already earning a retirement check, and because I am older than the rest of the staff.

During my years as a teenager and young adult, there were times and circumstances when I experienced racial discrimination because of the times. Even during my early years in the U.S. Army, racial discrimination did not surprise me. However, I learned to use those negative experiences to my advantage. It was important to me to learn how to deal with discrimination without letting it get into my head.

I must tell you that what happened to me as the Nashville Recruiting Company First Sergeant affected me in ways that I cannot explain. I never envisioned racial discrimination would happen to me during that stage of my career, especially after all I had seen and done while serving in the U.S. Army.

I was the soldier who never discriminated against anyone and dared anyone who worked under my supervision to treat others differently because of personal dislikes for any reason. I absolutely despise discrimination in any form.

What the leadership of the Nashville Recruiting Battalion did to me was disrespectful to me and to other soldiers who have worked hard to complete the mission of the U.S. Army. It was also a disgrace to be removed from a position without explanation and denied the promotion to Sergeant Major that I overwhelmingly deserved, a promotion that I should have received, at the latest, the year I graduated from the United States Army Sergeants Major Academy, just like other senior soldiers who graduated from the academy.

In addition, I was a soldier who graduated with a Superior Rating in two of the three subject matters. Why should I be treated differently from students who did not pass any of the tests that were administered but were promoted during or immediately after completing the course? Don't get me wrong, I don't have a problem with anyone being promoted after completing such a difficult course, whether they passed or failed the test. No written test will measure a person's ability to perform on the job. To me, the Academy was there to enhance the abilities you already possess to be a Sergeant Major in the United States Army.

Why didn't I get promoted to Sergeant Major? I blame the Commanding General, who was newly assigned as Commander of the Military Personnel Records Center from the Recruiting Command. Although I don't have proof of this, I feel strongly that he pulled my name from the list just because he could.

Why the Commanding General pulled my name from the list remains a mystery to me.

My mind goes back to the expression on the face of my then Company Commander when he walked into my office and stated out of the blue, "First Sergeant, you know who is going to be the Commander of the Military Personnel Records Center?" He went on to name the Commander of the Recruiting Command at that time and laughed. I knew the captain was alluding to the fact that I would never be advanced to Sergeant Major. However, for the life of me, I did not believe it would happen. I did not think that a U.S. Army Commander would stoop so low as to lose his integrity by taking premeditated action against a senior Noncommissioned Officer based on what a subordinate told him without checking it out first. That's totally irresponsible for any person in a leadership or command position.

It has been over 30 years since I was discriminated against as a First Sergeant without an explanation. I am now 80 years old, and it still hurts me that those in power were allowed to discriminate against me, simply because I am Black. Those in power took from me the opportunity to lead soldiers as a Sergeant Major and perhaps a Command Sergeant Major in the United States Army, just because they could.

Here is what I believe. I believe that the Nashville Battalion Management was saving the First Sergeant position for a white person. Because I was forced into the Battalion, they did everything they could to remove me as quickly as possible. Hell, I was not even there long enough to learn the layout or directions to the airport or each of my recruiting stations without checking the map first. It is irresponsible for any Commander or leader to remove any soldier in a leadership position without giving them the opportunity to succeed.

It was an absolute honor for me to serve my country in the United States Army. I took pride in wearing my uniform and found pleasure and excitement in continually accomplishing the missions assigned to me. I am an absolute patriot. I remain loyal to the United States Army, the United States Military, and the United States of America.

To this day, I get emotional when our military men and women participate in openings at Tennessee Titans football games. To this day, I find myself getting emotional when I hear the words and music of our National Anthem. To this day, I get emotional each time I hear the bugler play taps at our veterans' Celebration of Life.

I fly my American flag at my home every day and follow the lead from our state and our president for the temporary positioning of the flag as we honor deserving dignitaries. I salute those military personnel I see doing great things for our country in any capacity. I grieve for the families of those soldiers who paid the ultimate sacrifice for our freedom during their service to our country.

I end with this very true quote in the words of John Robert Lewis, U.S. House of Representatives:

"All of us, it doesn't matter if we are Black or White, Latino, Asian American, or Native American, it does not matter if we are straight or gay. We are one people. We are one family. We are one house. We all live in the same house."

I say to you, "I Am an American."

Henry Clarence Wilson Sr.